THE AMATEUR
HORSE BREEDER

PELHAM HORSEMASTER SERIES

THE AMATEUR HORSE BREEDER

Revised Edition

A. C. Leighton Hardman

PELHAM BOOKS

First published in Great Britain by
Pelham Books Ltd
44 Bedford Square
London WC1B 3DP

First edition published 1970
Reprinted 1971, 1972 and 1973
Second edition published 1974
Third edition published 1980
Fourth edition published 1985
Reprinted 1985

ISBN 0 7207 1607 1

Typeset by Cambrian Typesetters, Farnborough
Printed in Great Britain by
Hollen Street Press, Slough

To my parents and brother

CONTENTS

ACKNOWLEDGMENTS

I should like to express my gratitude to Dr Leo B. Jeffcott, Ph.D., F.R.C.V.S. for his helpfulness and kindness to me in reading and correcting this book, also to the veterinary surgeons at Reynolds, Leader, Day and Crowhurst, Newmarket for their invaluable advice. To Mrs P.J. Leader for putting up with the noise of a typewriter. I am also grateful to Ann and Tina for all their help, especially to Ann for the unlimited loan of her typewriter.

I would especially like to thank Sidney Ricketts, F.R.C.V.S., for reading and correcting this new revised edition, and Anne Hammond for her helpful suggestions.

A.C.L.H.
Newmarket

FOREWORD TO THE FIRST EDITION

by Major-General Sir Evelyn Fanshawe, C.B., C.B.E., D.L.

The author of this book though young in years has made a very thorough study of her subject and what is more has taken great trouble to get practical experience in breeding horses and ponies.

The subject of horse breeding can be matter for many volumes but this book is a sound and handy guide for the small breeder.

The opening paragraph is very true and puts in a nut shell why most of us want to breed a horse.

This book is written in simple language: the subject matter is thoroughly sound and I recommend it to all breeders and particularly to those who are just starting.

FOREWORD TO THE THIRD EDITION

by Dermot J. Forde, M.V.B., M.R.C.V.S.
Breeding Manager, Bord na gCapall

Over the past few years Ann Leighton Hardman has made a major contribution to the horse-breeding industry by producing books such as this one for the benefit and guidance of those directly involved in one way or another with the noblest of all animals.

Since first coming across the original edition I have recommended many would-be breeders to read *The Amateur Horse Breeder* and most of them, now busily engaged in producing horses and ponies of different breeds and types, use it as their standard work of reference.

From the most important task of choosing a mare with good breeding potential to start out with, to the sale of the eventual product, the reader is safely and pleasantly steered through all the various phases which may seem routine to the experienced breeder but which could be so inhibiting to the newcomer. It seems to me that nothing of a practical nature has been omitted and, in an era bedevilled with numerous and sometimes puzzling equine diseases, it is good to see such calm and sound advice as that given by the author on subjects such as contagious equine metritis and equine rhinopneumonitis. The stress laid on the association between good hygiene and disease control is also to be welcomed.

When all is said and done, however, it is the attention paid to the basic principles and to detail which will determine whether or not a breeding enterprise will be successful. In this respect Miss Leighton Hardman has done her homework most diligently and the information on breeding procedures, feeding, worming, foot care, and the proper preparation of a horse for show or for sale should be taken to heart by everyone.

1 Your Mare

A foal of your own: something you have bred yourself from your own mare, by a stallion which was entirely your own choice, halter broken and handled by no-one but you – in fact, all your own work; this is the dream and ambition of many horse owners but unfortunately realised by only a few.

Perhaps one of the main reasons why more people do not breed their own foals is just lack of knowledge, as this is a subject outside the realm of everyday horse management but one in which anyone with the know-how can become really interested, as the pride of owner-ship in a bought animal is nothing compared with that felt by the breeder.

When deciding whether or not to breed a foal from your mare one of the most important considerations is her age. It is unwise to get a mare in-foal before she is at least three years old. Horses continue to grow and develop until they are about five years old, but up to the age of three years the rate of development is very fast and to get a mare in foal at this critical period in her life could stunt her for ever.

Some authorities state that the most difficult age to get a mare in foal is at three years, though from my own experience I have not found this to be particularly so. However, mares over fifteen years old tend to become less fertile with increasing age but if kept breeding regu-larly will often go on producing foals into their twenties.

There are on record instances of mares producing their first foals in old age, as long as a mare comes in season regularly and puts up a good egg, there is absolutely no reason why she should not get in foal, even if she is twenty or more years old. Likewise, there is no real reason why she should have a difficult foaling. Unlike

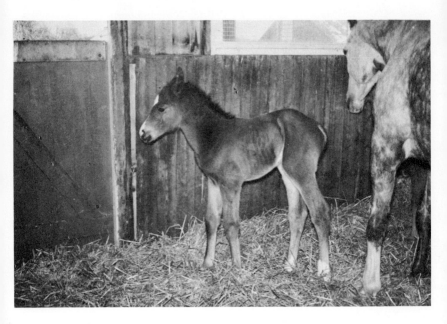

1. A foal of your own.

her human counterpart, a mare can remain fertile until
she dies of old age.

An equally important consideration is the conformation and condition of the mare's internal and external
genital organs. We will deal first with the external organs,
as these can be checked by the owner himself, whereas
only a qualified veterinary surgeon can check the
internal organs satisfactorily.

First of all look at the mare's udder: make sure that it
is free from warts and that she has two teats of even size.
Next, lift her tail and compare the conformation of her
vulva with the diagram in Fig. 1. A mare with a sloping
vulva is seldom a good breeding proposition in the
natural state. At times, most especially when in season,
she will be heard to draw in air through her vulva. This
air passes down into the uterus where it can set up
microbiological infections which in turn can lead to

sterility, or loss or resorbtion of a foetus. This is commonly known as 'wind-sucking' — not to be confused with the vice of the same name.

This malformation of the vulva can, however, be rectified by a simple operation (known as Caslick's operation) carried out by your veterinary surgeon. Using a local anaesthetic he simply cuts down the sides of the vulva and stitches these parts together down to the brim of the pelvis; this prevents the lips of the vulva from hanging open loosely at the top end and enabling air and dung particles to enter the vagina. The stitches are removed in about ten to fourteen days. This operation does not in any way interfere with the mare's ability to urinate freely nor, unless she has been stitched down a long way, with subsequent service by a stallion (though if this is so the bottom stitch is removed and replaced after service).

If your mare happens to be a grey, turning or already turned to white, you may notice, on looking under her tail and in the region of her vulva, a collection of small hard black lumps, sometimes known as melanoma or 'black cancer'. These are tumours which, due to the loss of coat colour with advancing age, become heavily

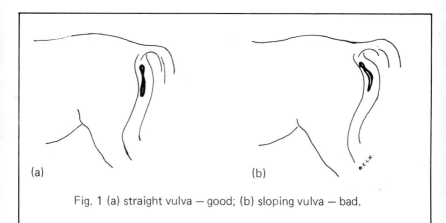

Fig. 1 (a) straight vulva — good; (b) sloping vulva — bad.

laden with the black pigment — melanin. Generally speaking they are nothing to worry about and very seldom interfere in any way with foaling. It is, however, a good idea to have a mare with these lumps examined by a veterinary surgeon before getting her served just to make sure that she will be all right.

A mare should come in season regularly once every three weeks and stay in season on average from three to nine days. If you do not know whether she comes in season regularly or indeed if at all, it is probably better that you should have her checked internally by a good horse veterinary surgeon before sending her to stud. If a mare never comes in season, there is always the possibility that she may already be in foal, especially if you have only just bought her and aren't familiar with her background. So you might be wasting your money and the stud's time by sending the mare to them before you have had her examined.

When deciding that you want a foal from your mare, make up your mind quite definitely what type of horse or pony you wish to breed. Stand back and look at your mare critically; if you can only see good in her ask a knowledgeable friend to help you. If she has any out-standingly bad points, choose a stallion which is particu-larly good in these respects. Each parent contributes fifty per cent to the genetical make-up of the offspring, so it is not always possible to get rid of a particular characteristic in the first generation, though an ex-ceptionally prepotent stallion will help.

Generations of unsound stock, which go lame im-mediately they are put into work, can be and indeed are produced when unsound stallions and mares are used for breeding. Unsoundness in this case does not include accidental injury, which can never be passed on to the resulting foal but only relates to inherited conditions such as those listed on page 23 and to poor conforma-tion.

It sometimes happens that an unsound mare which is

suffering from one of these listed conditions or has poor conformation, is offered free to a good home for breeding purposes. To safeguard the soundness of the horse for future generations, these mares should never be bred from, however tempting the offer may seem. Very lame mares may have difficulty carrying a big foal to term.

Regarding size: the first foal is usually the smallest and where there is any great discrepancy in size between the parents, the foal will, at birth, more often follow nearer to the size of the mare than to the stallion, this being nature's safeguard for an easy foaling.

When considering half-bred and pony mares of mixed breeding, better results will be obtained by sending them to pure-bred stallions. When using stallions of mixed parentage, however good, on mares of unknown or mixed breeding, any one of a vast number of characteristics may emerge, possibly with disastrous results. A stallion of pure ancestry tends by his many generations of pure breeding to be more prepotent, i.e. able to pass on his own characteristics in preference to those of his mate.

To obtain more size and quality in the first generation there is nothing better than an Arab or Thoroughbred sire. The Arab, due to his many centuries of pure desert breeding, is the most prepotent breed of horse today. Thus, he is the most capable of obtaining maximum quality from half-bred mares in the first generation. He is closely followed in this respect by the Thoroughbred, which can also impart extra height, so essential in hunter breeding. The mare, in turn, if she has substance, can help to produce bone, i.e the circumference of the leg immediately below the knee. A good hunter should have size, quality, and at least nine inches of bone; an ideal hunter sire should, therefore, have both size, quality and plenty of good flat bone.

During the summer before you intend to breed from your mare, you can occupy yourself profitably by attending some of the larger shows which have classes for hunter or pony youngstock. Observe not only the prize-winners but also the losers; note how both are bred. A lot can be learnt from other people's successes and mistakes. Pay special attention to the mare and foal classes. Look for mares with similar breeding and con-formation to your own and note the foals. If a foal has poor conformation, avoid that particular cross with your own mare. If a foal is a good individual, look for others bred similarly and if the result is equally pleasing, consider using that breed or cross yourself.

Before you finally decide to send your mare away to stud, consideration should be given to the amount of accommodation available. You will need double the grazing area, i.e. a minimum of four acres for every mare and offspring for all-year-round grazing and double the loose-box space. Initially, there will be little change (one to two acres is sufficient) but in a matter of about four months, the foal will be eating a considerable quantity of grass each day and by the time it is only six months old, its daily requirements will be almost identical to those of the mare. Therefore, it is unwise to breed from a mare unless you are sure you have the room to deal with *two* horses.

Thought should also be given to the foal's future. Unless you do not mind breeding horses for the meat trade, make sure you can keep a young horse, if necessary, until it is old enough to be broken and ridden — i.e. four years old.

When buying potential brood mares ensure that you receive *all* the relevant documents *at the time of purchase.* Without these, foals cannot be registered. It is also essential that changes in ownership of breeding stock are recorded with the breed society concerned.

2 Choice of Stallion and Stud

When you have finally chosen the type or breed of stallion you wish to use, the next thing to do is to decide how much you intend to spend on the stud fee and then look around for a suitable horse.

It is as well to remember that it costs as much to rear a bad foal as it does a good one and fifty pounds extra spent at this stage in travelling expenses or stud fee to a good stallion could make £500 difference to the value of the resulting foal. So don't use a local stallion just because he stands locally. If, on the other hand, your local stallion is also the best, then count yourself lucky and by all means send your mare to him.

There are several ways of finding out which stallions are standing at stud and where they are located. One of the best ways is to write to the breed societies concerned — most keep lists of stallions standing at stud in the British Isles and will gladly send you a copy if you write to them. Names and addresses of secretaries can usually be obtained from the British Horse Society (see p. 183).

Other useful methods of selection are to buy copies of *Horse and Hound* any time from November to May — this magazine carries an extensive range of stallion advertisements which embrace most breeds and types. Ask your farrier if he knows of any stallions at stud in your area or buy a local farming paper, as most of these contain stallion advertisements from after Christmas until early May.

Using any or all of these methods, compile a list of stallions in which you are genuinely interested and write to the studs concerned, asking for the following:

1 *A stud card*. This should give you a photograph of the horse and details of his breeding, winnings and winning progeny, if any; also his stud fee. With regard to the stud fee, this is normally divided into five categories:

(a)   Fee without any concessions – this must be paid regardless of whether the mare is in foal or not.

(b)   No foal, free return – this is self-explanatory but the free return must be taken the following year, usually with the same mare.

(c)   No foal, no fee – in this case the mare must be tested barren by October 1st otherwise the fee is due.

(d)   No live foal, no fee – the stud fee is usually payable at the end of the stud season unless the mare is tested barren but the fee is returnable in full, if the mare does not produce a live foal which is viable at forty-eight hours. Sometimes a stallion is advertised as first a live foal then payment of the stud fee.

(e)   Split-fee concession – this is a combination of (a) and (c). A percentage (usually half) the stud fee is due either on signing the contract or at the end of the stud season (July 15th). The balance becomes due on October 1st unless the mare is tested barren by that date.

Stallions standing at no foal, no fee or with live foal concessions usually have slightly higher stud fees, to compensate for the loss of income from mares which fail to conceive – the national foaling average for Thorough-bred mares is only two foals for every three years at stud. The stallion owner also reserves the right to refuse mares he considers unsuitable for any reason – this is stated as 'approved mares only'.

However, mares visiting stallions which do not offer any concessions can be covered by insurance (with Lloyds) before they go to stud, against loss of stud fee and expenses incurred whilst at stud. This can be extended to include the unborn foal, usually through to thirty days after birth, or for a longer period if required. It covers the insured against the mare resorbing or aborting a foetus, birth of a non-viable foal or a sick foal dying within the period of insurance.

2 *Fertility percentage.* This is the number of mares in foal in a given year compared with the total number

of mares served by the stallion. Ideally this should be at least seventy per cent.

3 *Ministry of Agriculture Stallion Licence**. It is as well to ask if the stallion is licensed, as all stallions of five years old and over, other than Shetlands and those registered in the General Stud Book, which stand at public stud, and all travelling stallions must have a Ministry licence. This means that they have been passed by a veterinary surgeon as being of good conformation and free from the following heredi-tary conditions:

cataract	shivering
defective genital organs	roaring and whistling
navicular disease	sidebones
ringbone (high and low)	bone spavin
stringhalt	parrot mouth

It should be noted that any mare used for breeding should also be free from these defects as she would be just as likely as the stallion to pass them on to her foal.

Each year in March a number of Thoroughbred stallions, registered in the General Stud Book and in-cluding many which have won races, compete for premiums at the Stallion Show in Newmarket. These premiums are subsidised by an annual grant from the Horse Race Betting Levy Board and enable the stallions to cover half-bred mares at a very reduced fee in dif-ferent districts of England, Wales and Scotland, to which they are allocated by the Hunters' Improvement and National Light Horse Breeding Society.

These stallions have all passed a rigorous veterinary examination and are therefore guaranteed not to possess any hereditary unsoundness which they could pass on to their offspring.

*The Minister has informed Parliament that he intends to introduce legislation to repeal the Horse Breeding Act as soon as is practicable. When this takes place all stallion licences will cease.

For the small breeder wishing to produce National Hunt horses, eventers, hunters or hacks from their mares, these stallions represent the best value for money in this country.

Many of these stallions have produced top-class National Hunt horses, including the following Grand National and Cheltenham winners: Highland Wedding, Pas Seul, Merryman II and Specify. Many top show hunters, both in-hand and under saddle, are also by these stallions, as are some very well-known three-day event horses and show-jumpers.

If you do decide to use a Hunters' Improvement Society premium stallion on your mare, besides having a low stud fee to pay, your mare will also be eligible to compete at certain shows for a premium. These are offered to mares which have produced a foal by a stallion awarded a premium under the Society's scheme in the year of covering. Premiums of £15 each are allotted. In all cases the mare must have her own foal at foot and no mare can take more than one premium in any year. Judges do not award premiums to mares considered below the required standard. The winning mare must be entered in the Hunter Stud Book, and her owner must be a member of the Hunters' Improvement Society before payment of the premium is made. The premiums are paid direct to the successful exhibitor by the Hunters' Improvement Society.

In Ireland horse breeding is subsidised by an annual government grant controlled by Bord na gCapall (Irish Horse Board). To conform with EEC regulations, the Bord organises the registration of non-Thoroughbred breeding stock. They inspect and register suitable stallions, mares and fillies and issue passports to those they pass. The passports contain a markings certificate and remain with the animal for the rest of its life.

Since Thoroughbreds are already well documented by Weatherbys, the Bord's concern is with the Irish Draught, Connemara pony and half-breds. Animals registered with a breed society have both documents incorporated into the one passport.

Details of other schemes and grants available to Irish breeders can be obtained by writing to Bord na gCapall (see page 183).

Control of genital infections. Ask the stud what precautions they normally take against the spread of genital infections. Enquire whether the stallion is swabbed annually for CEM and other bacteria, and what tests your mare will need before she is accepted onto the stud. A negative answer may save money for swabs in the short-term but could easily result in an infertile, infected mare or the birth of a dead or dying foal. Studs which have no system for disease control should be avoided.

Charges for keep. On Thoroughbred studs these are usually divided into three categories: foaling mares, maiden mares, and barren mares — i.e. those about to foal or with foals at foot, those which have never been served before and those which have had at least one foal but are without foal at the time of going to stud. There is another class of mare known as a barren-maiden: this is a mare which has been served before but which has never had a foal. Foaling mares are charged the most and barren mares the least.

Mares sent to a public stud to be covered are known as visiting mares. Unless instructed to the contrary, barren and maiden mares are kept at grass once the weather is warm enough for them to live out and the grass is growing. Many studs operate a scheme whereby barren stabled mares are charged a higher rate for keep than those at grass, so the choice can rest with the owner. Foaling mares are normally housed at night, at least until the foals are over a month old and the weather is settled. Some studs, however, prefer to have all the foals inside after dark. All visiting mares have their feet trimmed once a month by the blacksmith, as do foals over one month old. Worming takes place when the mares arrive (unless they have been treated recently or are close to foaling) and then at regular monthly intervals. Foals are normally dosed at about eight weeks old.

Keep charges vary enormously and can range from as little as £7 per week at grass on a pony stud to about £70 per week for a foaling mare on some Thoroughbred studs.

Establishments making very low charges for keep should be avoided. Studs must make a profit to help offset their labour and feeding expenses for the rest of the year. Low charges often mean that your mare will be returned to you much thinner than when she went to stud.

A groom's fee is usually charged and varies from £1 to £20 per mare, according to the stud fee. Any veterinary charges are sent to you direct by the veterinary surgeon attending the stud. Night duty, worming, shoeing and transport expenses are charged as extras.

Mares usually stay at stud for at least eight weeks, so a rough estimate of the total charges likely to be incurred at a pony stud are as follows:

	£
Stud fee	50.00
Groom's fee	2.00
Keep at £7 per week	56.00
Blacksmith	6.00
Worming	5.00
	£119.00 (plus VAT)

These charges are correspondingly higher if a mare goes to a Thoroughbred stallion, other than a Hunters' Improvement Society Premium Stallion whose fees for half-bred mares are similar to those set out above: i.e. £38 plus VAT for members, or £65 plus VAT for non-members.

Once you have sorted through the stud cards you receive, you will have an idea of the stallions in which you are still really interested. The next step is to telephone the studs concerned and make an appointment to see the stallions; it is very necessary to telephone the stud manager before you go, as most studs are extremely busy places during the breeding season.

Public studs are run by a manager, on the owner's behalf, or by the owner himself. The manager will sell you the nomination to the stallion, that is, the right to send a mare to the horse in that particular year. He will also tell you the cost, or the 'stud fee' as it is known. Once your mare is at the stud, the manager will keep you informed about her progress. Accounts are normally rendered monthly.

The stud groom is appointed by the manager and is responsible for all horse operations, including foaling, trying and covering. Like the manager he keeps detailed records of all the mares. Nowadays it is common practice for the two jobs to be done by the same person, namely a stud groom/manager. A very experienced groom is normally employed to look after the stallion, and is known as the 'stallion man'. On small studs this task may be included in the stud groom's duties. The stud groom directs the stud hands and students who are under his care. If you wish to telephone the stud or write to enquire about your mare while she is at stud, contact either the manager or stud groom for information.

When you visit the stud first look at the stallion in his box. Apart from his conformation pay special attention to his temperament. This is most important if you are going to look after and handle the resulting foal yourself, as bad temper or a mean disposition can be inherited. Next ask to see him out and have the groom trot the horse up for you — make sure his action is straight and that he moves well. Ask the owner or stud groom if you can see some of the stallion's progeny and if possible their mothers — this should give you some idea of the type of foal you can expect from your own mare. If your mare is in foal, it is as well to ask the stud groom whether the foals are handled and led each day.

If not, be prepared for a big wild foal by the time he is two or more months old and your mare has been tested in foal again.

While you are looking round the stud check the condition of the animals, the type of fencing, the state of the paddocks and the loose-boxes, i.e. are they clean, tidy and in good repair and run to the same (or a higher) standard as your own stable at home.

It is probably very difficult to see your own yard as other people see it but the first impression of any stud should be one that is clean and well swept, with the boxes in good structural repair, painted or creosoted and obviously mucked out every day. There should be no broken windows and all windows should have guards over them. There should be no nails sticking out in either the boxes or paddock fencing, and no holes in the boxes, especially low down. All doors must reach ground level – on lying down a horse could put a leg through a gap under the bottom of a door and get badly trapped. The water bucket or water bowl in the boxes must be cleaned out every day, as should the field troughs in the paddocks. It goes without saying that there should be no barbed wire on any well-run stud and there must be no unguarded open ditches into which foals or even older horses could fall. Mares with foals at foot should be kept in separate paddocks from those without foals, to prevent accidents.

These points are very important. It is too late when an accident has occurred for no stud accepts responsibility for your mare from the point of view of accidents or disease, although naturally every care is taken. The initial responsibility for selection of a public stud rests with the mare owner.

It is equally important that your own private stud should be maintained to an adequately high standard – otherwise avoidable accidents will occur, ending in total losses or reduced values of stock from injury. Young-stock are notoriously more prone to accidents than older

No. . . . .

. . . .

. . . .

. . . .

. . . . 19 . . . .

Dear . . . .

I agree to take one nomination to

. . . .

for season 19. . . . . .on the following terms:

Stud fee. . . . . .Groom's fee. . . . . . .

Keep of Mare. . . . .

Yours faithfully,

Details of mare sent to stud:

Name:. . . . .

Sire: . . . . .Dam:. . . . .

Age: . . . . .Height: . . . .

Number of previous foals:. . . . .

Was she served last season? . . . . .Date:. .

Is mare to be sent: in foal, with foal at foot or barren? . . . .

Is the mare stitched?. . . . .

Date last wormed?. . . . .

List diseases (if any) against which the mare has been immunised?. . . . .

To:

. . . . .Stud

(Every precaution will be taken against accidents or disease, but no liability can be accepted therefore.)

Mares must not be sent shod.

Fig. 2. Example of a simple nomination form.

horses, so greater care needs to be taken when establishing a stud, no matter how small.

When you have finished looking at the stallion and have been shown round the stud, it is always very much appreciated if you give the stallion man or stud groom a tip, especially if you intend to send your mare to the horse.

When you have visited all the studs on your list you will be in a position to decide exactly which stallion you intend to use. Write to the stud of your choice and ask them to send you a nomination form. This is a contract which should be signed by every mare owner buying a nomination to a stallion. These forms are usually in duplicate — one copy being retained by the mare owner, and the other by the stud. Most studs also require details of the mare to be sent, and it is most helpful if you can supply them with as much information as possible. It is as well to book your nomination early in the year, as most of the best stallions fill up quickly — an adult stallion normally takes about forty mares each season.

3 Sending Your Mare to Stud

The stud season officially starts on February 15th and ends on July 15th, but due to the fact that many ponies are foaled outside, their owners like them to foal in warm weather so the season for pony studs extends into late summer.

The gestation period for mares is eleven months, or 340 days, although mares can foal a fortnight early and some a month late (see Figs 3 and 4.)

As previously stated mares come in season approximately every three weeks, therefore a mare served by a stallion but not holding to that service will usually come in season again three weeks later. These points must be taken into consideration when fixing a suitable date for sending a mare to stud.

If you intend to show your foal, as a foal or yearling, an early one, i.e. born in February, March or at the latest early April, is essential — the larger the foal the more likely it is to win a prize, always assuming its conformation is correct. Even a few weeks makes a big difference to size during the first year.

Make sure your mare is healthy and in good condition before sending her to stud. Over-fat mares are notoriously difficult to get in foal, so if your mare has put on weight, a course of slimming before Christmas, followed by hard feeding some six weeks before you want her to conceive, will assist the stud in getting your mare in foal. Thin mares in improving condition are usually easy to get in foal providing they are in good health and not thin due to worms or some obscure illness. Once in foal they will often put on flesh rapidly — this is probably due to an improvement in the ability to assimilate food which appears to take place once a mare is in foal. At the end of each stud season any mare found to be barren should be swab tested the

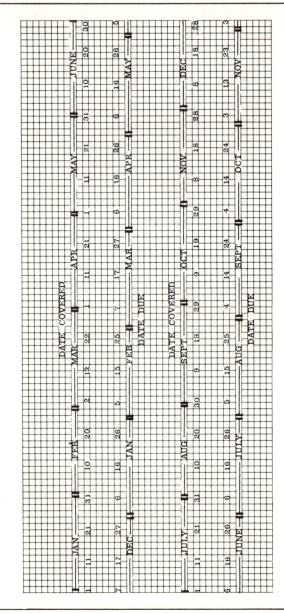

Fig. 3. Gestation chart. Aborted foals are those born forty days or more before their due date; they are unlikely to survive. Premature foals are those born between forty and fifteen days early.

first time she is seen in season. If she has a genital infection this can then be cleared up by your veterinary surgeon during the winter, thus saving valuable time during the breeding season and so enabling the mare to have an early foal the following year.

At least one month before you intend to send your mare, write to the stud and ask for details of any negative swab test results (e.g. for CEM, Klebsiella and Pseudomonas) they may require before accepting the mare.

The swab test is carried out very simply by the veterinary surgeon while the mare is in season and the cervix is open. He inserts a speculum (a long hollow tube) into the mare's vagina, down which he passes a sterile swab on the end of a wire rod which is contained in a narrow-bore metal tube. The swab is rotated gently within the cervix and withdrawn. A clitoral swab is also taken when examining the mare for CEM. The swab is plated out in a laboratory and after incubation the resulting growth on the plates is examined for pathogenic organisms. If a mare is infected a course of antibiotic irrigations while she is in season will usually put her right.

A few days before you actually send your mare to stud worm her – this helps to prevent the stud's paddocks becoming horse sick. Also, make sure you have had your mare's hind shoes removed. All studs make this a condition before they allow the stallion to serve a mare or turn her out with the other mares as untold damage can be done by a shod horse.

Notify the stud as soon as possible of the exact day and time you intend to arrive with your mare. This is best done over the telephone, so that a time suitable to both parties can be arranged.

When taking your mare to the stud for any length of time make sure she is wearing a leather headcollar which is exactly her correct size and bearing a brass plate engraved with her name. In this way the headcollar will not be lost and the staff will be able to learn your mare's name easily and recognise her in the paddock.

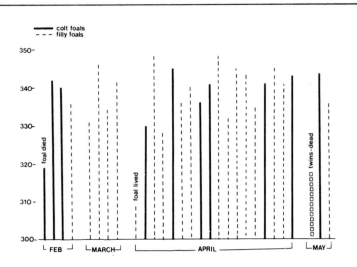

Fig. 4. Random sample of mares showing length of gestation.

Mares can conceive only if they are served while in season. The visible signs that a mare is in season are: she will call to other horses in the neighbourhood; if she is out at grass, on seeing another horse she will often rush up to it and show very marked signs of interest; on coming into contact with other horses she may lean towards or against them, raise her tail and pass water; her vulva will be relaxed and will continually open and close which is known as 'winking'; if the lips of the vulva are parted the mucous membrane inside will be seen to be moist and red in colour as opposed to the usually dry, salmon pink of a mare not in season.

The mare has two ovaries which are bean shaped and about the size of a small hen's egg. Each is suspended by a ligament from the roof of the abdominal cavity about an armslength from the vulva. Each ovary has several raised areas on its surface, one being generally much larger than the others. These elevations are known as Graffian follicles which gradually ripen and in turn discharge a ripe egg (ovum), the process being known as ovulation. The egg is discharged at the point on the ovary known as the ovulation fossa, which occurs in the middle of the concave side. As the follicle ripens so it becomes larger and softer, usually measuring from 2.5cm to sometimes as much as 7cm across before ovulation.

If the mare does not get in foal, as one follicle ruptures another one starts to enlarge. After the rupture of a follicle a small round scar is left in the ovary known as the yellow body or corpus luteum. This gradually gets smaller but where fertilisation has occurred the yellow body persists throughout pregnancy. During pregnancy other follicles rupture and more yellow bodies are formed. The corpus luteum is in itself an endocrine gland producing the hormone progesterone which is essential to the maintenance of pregnancy. Follicles differ from the corpus luteum in that they contain fluid. Eggs escape from a ruptured follicle about every twenty-one days and pass down the oviduct and fallopian tube where fertilisation should take place. If fertilisation does not occur, the egg remains in the fallopian tube and is gradually resorbed; only fertilised eggs enter the uterus. Ovulation usually occurs twenty-four hours before the mare stops being in season. For successful fertilisation,

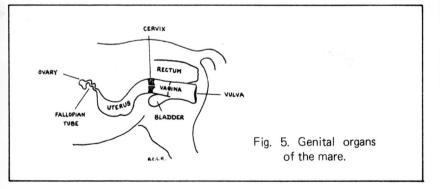

Fig. 5. Genital organs of the mare.

live sperm should be present at the moment of ovulation as the egg dies twelve to twenty four hours after being shed and as fertilisation must take place in the oviduct or fallopian tube, live sperm must therefore be present in the tube at this time. The fertilised egg then passes on down into the uterus and there pregnancy commences.

During the winter months, the ovaries are often small and inactive with no evidence of follicles or corpora lutea and there is a low progesterone level in the blood. To save unnecessary expense it is advisable to test barren and maiden mares to determine their blood progesterone level, and not send them away to stud until they have had a progesterone reading. To speed the on-set of a normal spring cycle most authorities recommend leaving the light on in a mare's loose-box after dark, long enough to increase the hours of 'day-light' to sixteen hours. Two-hundred watt incandescent bulbs must be used for this purpose.

Sometimes, especially in February and early March, when a mare is in season the follicle will not ripen but remains on the ovary or the mare will continue to put up follicles without forming a corpus luteum; either way she will stay in season for some considerable time, sometimes for a month or more. In this state the mare cannot be got in foal however many times she is covered. In this case, the mare will have a progesterone reading and will normally respond to treatment with the drug Regumate.

Mares with a persistent corpus luteum will not come in season at all. If a blood test shows a high level of progesterone, things can be rectified by injecting the mare with the hormone prostaglandin; this gets rid of the corpus luteum, allowing the mare to develop a follicle and so come in season.

It is almost essential to leave your mare at the stud at least until she is no longer in season. Most studs serve their mares twice before they go off – this is due to the fact that although sperm from a very fertile stallion will live for as long as a week under ideal conditions inside a mare, under average conditions they will probably only live about three days and in older mares they sometimes only survive for twenty-four hours. As we have seen, the egg is not liberated until the last twenty-four hours before the mare goes off and in order that it has the maximum chance of being fertilised active sperm must be present at the time of ovulation.

When your mare arrives at the stud the stud groom will probably check her for signs of contagious disease, e.g. strangles, coughing, lice, etc. If your mare is in season he may check to see if she is showing signs of vaginal discharge. As infected mares don't always show signs of discharge, all well-run studs require mares visiting their stallion to be swab tested each time they are in season before service. This prevents any chance of the stallion becoming infected and thus passing on an infection to other clean mares.

All mares are tried every morning to ascertain which, if any, are in season, with the exception of: foaling mares – those due to foal; newly foaled mares; and those within ten days of being covered. Trying is the process whereby man, using a male horse for the purpose, determines the exact sexual state of mares at any given time. Since trying represents some danger for the stallion, many studs use an inferior stallion to try the mares; this horse is known as a 'teaser'.

The act of getting a mare in foal is known as covering or serving her, although the season is always referred to as the covering season or stud season. It is often possible to send a mare to a teaser to be covered. Most studs like their teasers to cover a few mares each year, as this maintains their interest in the other visiting mares. The service fee is normally a tip to the stud groom. The teaser's degree of inferiority usually relates to the value of the covering stallion. For instance, the good National Hunt stallion Question – sire of the Grand National winner Highland Wedding — worked as a teaser in New-market towards the end of his life.

The most common method of trying mares is to use a solid wooden board, known as a 'trying bar', to protect the teaser from the mares. Normally the board measures eight feet long by about five feet high; the exact height varies according to the size of animals being bred. It is normally padded with matting or rubber to help prevent injuries and often has a free-running roller bar along the top. The stallion and mares are led to opposite sides of the bar for teasing, or trying.

Trying is one of the most important operations of the day on a public stud; that is, unless the stallion is running with his mares under natural conditions. When the stallion and his mares are kept apart, it is entirely up to the stud groom to determine when the mares should be covered, therefore the stud groom is a key factor influencing a stallion's fertility.

The mares are led up to the teaser one by one and allowed to greet him. The stud groom stands behind the mares, out of harm's way, and makes a note in his book of each mare's reaction to the stallion. Mares not in season will usually lay back their ears, kick and squeal —

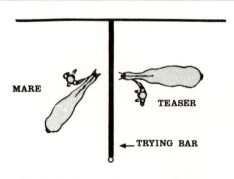

Fig. 6. Stallion and mare at trying bar.

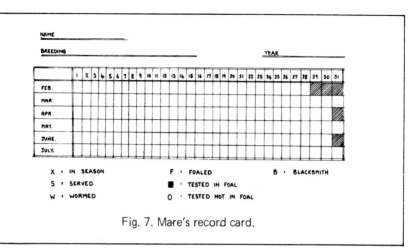

Fig. 7. Mare's record card.

hence the need for a trying bar. Mares in season will lean towards the horse, stand with their hind legs apart, lift their tails and pass water. The stud groom then transfers the information he has collected to a stallion wall chart, normally kept in the stud office. This chart is designed on similar lines to the record card shown in Fig. 7 and gives a picture of each mare's sexual cycle. Well-run studs have all visiting mares examined by their veterinary surgeon before covering. He takes a swab from the genital tract while the mare is in season and submits it to a laboratory for bacteriological examination. This safeguards the stallion and all his mares from transmission of venereal disease. The veterinary surgeon will also examine the mare to ascertain that she has only one ripe egg. Covering a mare with two eggs may lead to the birth or abortion of twins (but see page 47). Mares can only conceive if they ovulate, so covering a mare which has no sign of ovulating within the next few days is considered a waste of everyone's time.

Under normal circumstances stallions cover a maximum of two mares per day — one in the morning and the other in the afternoon — but at the height of the season they may cover three or four mares. Covering normally takes place in a purpose-built barn or yard, known as the 'covering yard' and situated close to the stallion unit. The floor is of a suitable non-slip material, usually bark, peat or sand. A slope or mound is constructed in the yard so that small mares can be placed uphill when covered by large stallions, and big mares positioned downhill to smaller stallions.

The mare is brought into the covering yard and prepared for service. Both the stallion and mare wear bridles for maximum control. Felt boots are normally put on the mare's hind feet to reduce the chance of damage being done should she decide to kick the stallion. Personnel wear disposable plastic gloves when handling the hindquarters of mares, to reduce the possibility of spreading venereal disease. Tape or a disposable bandage is applied to the mare's tail to keep long hairs out of the way. Cotton-wool swabs soaked with a mild disinfectant solution (two tablespoonfuls of Savlon liquid to a pint

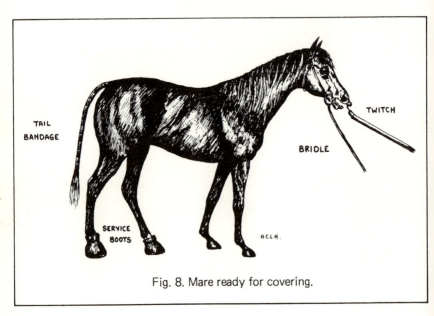

Fig. 8. Mare ready for covering.

of warm water) from a spray gun, are used to wash the mare's vulva and hindquarters before service – a fresh swab being used for the vulva and each side of her buttocks. The disinfectant is rinsed off in the same way with clean water. Buckets should not be used for washing mares' or stallions' genital areas as these can be a source of infection. Mares are normally twitched for covering not as a means of forcing service but merely to protect the stallion.

When the mare is ready, she may be tried again as a final check before covering. If she is a maiden, the teaser is often allowed to climb onto her back, in order to get her used to the idea, before covering by the stallion – this is known as 'bouncing'. If the mare stands quietly, she is covered. Mares still in season two days later are covered again, known as a 'cross service', and this is repeated until they have ovulated.

Most mares come in season about seven to ten days after foaling and then on a regular three-weekly cycle during the breeding season. However, some mares do not follow this normal behaviour pattern and may come in season at almost any time from about ten days after they were last in season. This is the reason why stud grooms normally try mares every day, and is the best reason why mares should be left at stud until tested in foal and not taken home after covering. Once the mare has gone forty days without coming in season, otherwise known as 'holding' to her service, she can be examined or tested for pregnancy, details of which appear on page 44. A mare which comes in season again after covering is said to have 'broken'.

After the mare has been covered and is no longer in season, 'gone off', you will receive a card similar to the one shown in Fig. 9 telling you that your mare was covered and the dates this took place. It is a very good idea to keep an accurate record of when your mare is in season and covered, each year. Copies can be drawn up of the chart on page 39 and used to record the in-

```
. . . . . . . . . . . . STUD

Dear
              Your mare. . . . . . . . . . . . . .was covered
by . . . . . . . . . . . . . . . . . .on . . . . . . . . . . . . . . . . . .
                  Yours faithfully,

                      Fig. 9.
```

formation. These charts would be of help to stud managers and could accompany the mare to stud each year.

The covering card (Fig. 9) should not be confused with the covering certificate issued at the end of the season to mare owners, whose mares visited Thoroughbred or Arabian stallions, for instance. This covering certificate must be kept safely and sent to the breed society concerned together with your annual returns when the time comes to register the foal. If in doubt, it is advisable to ask the stallion owner for details of foal registration.

Thoroughbred mares registered in the General Stud Book must have a return made on special forms supplied by Weatherby's for each year they are at stud. The return is made the year after covering, so that details of the foal can be given. The return must be made even if the mare is barren. Thoroughbred foals must be registered with Weatherby's by July 31st (in the year of birth) to qualify for the lower registration fee. After that date the fee goes up, with a further increase on October 7th and on December 31st. Many other breed societies have similar schemes. Thoroughbred foals must have their markings recorded by a veterinary surgeon before four months old, otherwise they will not be accepted into the General Stud Book.

It is sometimes possible to register a non-Thorough-bred mare and therefore her foals by Thoroughbred stallions in Weatherby's half-bred register when any of the following conditions hold good and the mare was foaled before 1974. The mare:

(1) has winning form Under Rules, i.e. Flat, National Hunt, or Point-to-Point;
(2) has bred a winner Under Rules;
(3) is closely related to a winner on the dam's side;
(4) is ineligible for the General Stud Book but registered with a turf authority outside the British Isles;
(5) has any other reasonable cause for being registered, e.g. is at least 15.2 h.h., and/or by a Thoroughbred sire, out of a mare also by a Thoroughbred sire.

Annual returns must be made for registered half-bred mares.

If a registered Thoroughbred or half-bred foal has dual parentage — that is, the mare was covered by two or more stallions in the same season, for instance when a stallion dies suddenly — Weatherby's will arrange for the foal to be blood-typed, so that the correct sire can be established. This is an accurate method of determining parentage.

Further details of your mare's life on a public stud can be found by reading *Stallion Management* (also published by Pelham Books).

4 Care of the Mare up to Foaling

It is a good idea to have a pregnancy test carried out on your mare before she leaves the stud. This way you will know definitely if she is in foal and then if she proves barren and the season is not too far advanced further steps can be taken to get her in foal. If, on the other hand, your mare is already at home and the stud season is over it is still a good idea to have her tested so that you will know how to deal with her: whether to hunt her or not, how to feed her and how much work to give her.

There are four tests available:

(i) *Ultrasound scanning for pregnancy diagnosis* — this is a new technique introduced and used extensively on studs in the major breeding areas since 1982. The principle relies upon the transmission and analysis of high-frequency sound waves. The sound beam penetrates tissues without harming the mare, foetus or operator. The pregnancy can be seen clearly as a dark, well-defined, solid circle, in one or other horn of the uterus. It can be detected as early as fourteen days, when it measures some 12–16 mm diameter. By about day twenty-five, the foetus can usually be seen as a small white spot within the black circle. By fifty-five to sixty days after covering, it is generally possible to recognise a tiny head, backbone and legs.

The scanner is therefore a very useful modern technique for early pregnancy diagnosis but more important still as a means of:

(a) Picking out mares which have conceived twins — especially those lying side by side in the same horn of the uterus.

(b) Detecting mares in prolonged dioestrus — those which although they have not conceived have not come in season again. A condition easily overcome with a single injection of prostaglandin.

(ii) *Manual test* — carried out by a veterinary surgeon at any time from forty days after the last service date, up to foaling.

(iii) *Blood test* — a blood sample (30 cc) is collected from the jugular vein by a veterinary surgeon between fifty and ninety days after service and is sent away for testing. (Optimum time seventy days.)

(iv) *Water test* — a sample of urine is collected by the owner any time from 120 days after service and is sent away by a veterinary surgeon for testing. A method of collecting a urine sample for the water test is as follows:

To obtain the best sample for testing this should be taken first thing in the morning after the mare has been left in a box over-night without water.

Equip yourself with a clean plastic bucket — a metal bucket is too noisy. Fill a second bucket with some warm water, rub plenty of soap on your hand and soap the inside of the vulva well — this sets up an irritation which causes the mare to urinate quickly. Remember to use disposable rubber gloves and so prevent spreading infection from one mare to the next. If possible take the mare to a different loose-box which has been occupied by another horse. Allow her to smell the bedding, shake some straw under her and whistle. The moment the mare shows any signs of staling, put the bucket under her tail. Collect enough to fill a medicine bottle, transfer the sample from the bucket to a suitable bottle, label this with your name, your mare's name, age, her last covering date, date of collection and your veterinary surgeon's name and address. Give the sample to your veterinary surgeon. It usually takes about one week for the results of both the blood and water tests to come back.

An alternative method for collecting a urine sample is to fix an aluminium or plastic jug to the end of a long

CODE NO. OF MARE	POSITION OF RIPE EGG AT TIME OF OVULATION	POSITION OF FOETUS AT TIME OF MANUAL PREGNANCY DIAGNOSIS
1	Left ovary	Right horn *
2	Left ovary	Right horn *
3	Right ovary	Left horn *
4	Right ovary	Right horn
5	Left ovary	Left horn
6	Right ovary	Right horn
7	Right ovary	Right horn
8	Left ovary	Left horn
9	Right ovary	Right horn
10	Right ovary	Left horn *
11	Left ovary	Right horn *
12	Right ovary	Right horn
13	Right ovary	Right horn
14	Left ovary	Left horn
15	Right ovary	Right horn
16	Left ovary	Left horn
17	Left ovary	Left horn
18	Right ovary	Right horn
19	Left ovary	Left horn
20	Left ovary	Left horn
21	Left ovary	Left horn
22	Left ovary	Left horn
23	Right ovary	Left horn *
24	Left ovary	Right horn *
25	Left ovary	Right horn *

Fig. 10. A random sample of in-foal mares showing the position of the foetus at forty days, related to the position of the ripe egg at service time. Those which have altered position (indicated by *) amount to thirty-two per cent.

stick. This apparatus can then be passed quietly behind the mare the moment she stales, without frightening her unduly and without any chance of your being kicked.

(It should perhaps be noted here that when a urine sample is required for bacteriological examination it should be collected from mid-stream of the urination *direct* into a sterile container which should be corked immediately with a sterile stopper.)

The accuracy of any of these tests lies solely with the person carrying them out.

Union between the mare and the foetus is not complete until about thirty-six days after service; up to this time the foetus is held unattached within the uterus: after thirty-six days the pregnancy becomes established and loss of the foetus is less likely. However, in some cases early foetal death and pregnancy resorption does occur, usually by day ninety. It is now thought that one cause may be antigenic compatability between the mare and stallion. Therefore many owners have their mares rechecked at nine and twelve weeks, especially in the early part of the season when there is still time to get the mare covered again if she proves to be no longer in foal.

With regard to twins, it used to be suggested that this be prevented in the early stages by manual examination of a mare when she came in season and by not covering if two eggs were put up — but this did not prevent the possibility of two eggs lying side by side or a second egg suddenly ripening and being fertilized and so producing twins. It is now known that only five per cent of mares with twin follicles actually conceive twins, so covering is normally recommended.

It does, however, sometimes happen that although twins are produced they are not both carried through to parturition but that at a very early stage one fails to develop naturally and is resorbed while the other is carried to full term. Thus when a veterinary surgeon on pregnancy examination discovers twins he usually rechecks the mare a week or so later providing this is before day thirty-six to determine if they are both developing normally; if this is the case and the owner so wishes it, the mare can be injected with prosta-glandin and a fresh start made or the twins can be left —

although twins are more often than not slipped before they reach full term. If diagnosed with the aid of the scanner at an early stage, one foetal sac may be manually ruptured, providing there is one sac in each horn. If both foetal sacs are in the same horn prostaglandin should be used.

Thus if a mare is carrying twins the owner has two courses of action open to him: (i) leave them alone and hope that the mare may resorb one foal; or (ii) have one of the foals removed by rupturing its foetal sac, or alternatively, with an injection of the hormone prostaglandin, given before the foetuses are thirty-six days old.

The main disadvantages of twins are:

(i) the great possibility that they will not be carried to full term and thus slipped or born dead;

(ii) if they are born alive they tend to be small and remain stunted all their lives. Although good feeding can help to make good the pre-natal low level of nutrition, it can never quite make up for the early deficiency;

(iii) the danger of damage to the mare during foaling and even possible loss of the mare and the risk of not being able to get her in foal again until the following year, also the risk of permanently lowered fertility as a result of carrying twins,

(iv) the possibility that the mare will not be able to rear two foals and the consequent problem of obtaining a foster mother for one of them or alternatively rearing it by hand.

About half way through the gestation period the foal can often be seen to kick inside the mare. This is particularly noticeable if the mare is left without water for a few hours and then given a bucket of cold water to drink.

Other good signs that the mare may be in foal are that she will often appear more docile and her underline will begin to drop noticeably. This can best be seen if photographs are taken of the mare at approximately monthly intervals and compared one with another, or her girth measured with a long flexible tape measure.

When you collect your mare from the stud it is probably better not to load her into a very narrow compartment where she is squeezed against the sides of the trailer or horse-box. At all costs avoid upsetting an in-foal mare and drive carefully. On the other hand travelling has been known to induce ovulation in newly covered mares, with happy results.

If the mare returns home soon after service do not ride her hard or in any way upset her too much; when she is safely in foal you can carry on as usual but the mare herself will gradually get slower as she gets more heavy; until towards the end you will probably find it easier to turn her out in the field than to ride her, although quiet exercise is essential for an in-foal mare right up to foaling.

If foaling mares cannot be turned out each day into a paddock for exercise, or if on being turned out they tend to stand around and not exercise themselves, they should, whenever possible, be led out in hand for about half an hour each day for the last month or two before foaling, when they are too heavily in foal to ride.

Exercise is of great importance where the brood mare is concerned. Lack of exercise just before foaling can lead to difficult foalings and retained meconium in the foal. Therefore, even in bad weather, mares should have some exercise each day. The only exception to this rule would be when the ground is covered with ice or snow, where the risk of a fall and probably abortion would far outweigh the lack of exercise, so the mare should be left in her loose-box.

If you have the facilities, a straw ring can be made for winter exercise. To do this, instead of taking all the wet straw and droppings to the 'muck' heap each day they can be spread thickly in a ring, round which the horses can be walked. Straw is not very safe when placed on ice as it tends to slip unless it is very thick. A good method of rendering concrete paths safe to walk a horse on in icy weather is to scatter salt, plenty of ashes or sand on them; this must, however, be renewed each day as it sinks into the ice as the latter melts and the surface becomes slippery again.

Brood mares should be wormed regularly, the last time being one month before foaling, as some foals will play with and lick or eat their mother's droppings, so this will reduce the risk of the foal picking up a worm infestation from its dam. Consult your veterinary surgeon as to the best drug to use near foaling time.

Once mares are brought in for the winter, they should be supplied with a salt lick, ad lib hay and water. Mares that are in foal will need more bedding than usual and must be kept bedded down all the time as they will often lie down during the day for a rest.

Mares in foal require a slightly more nutritious diet than barren mares. They should be fed three or four times a day, e.g. a medium dry feed in the morning, a small feed at mid-day and a large mash or dry feed at night.

All breeding stock should be vaccinated against flu and tetanus; brood mares should receive an annual booster injection about four weeks before foaling. In this way, the foal should receive protection through its mother's first milk or colostrum; this may give protection for the first two months, after which it can be vaccinated itself. It is, however, a wise precaution to inject the foal with tetanus serum at birth and again six weeks later.

A few months before the mare is due to foal you must make up your mind what to do with her. There are three possibilities:

   (i) send her to the stud to foal down;
  (ii) foal her yourself in a loose-box;
 (iii) let her foal outside in a field.

Method (i) is probably the best if you have had no

previous experience of foaling and intend to breed from the mare again that year. Most studs are equipped with special foaling boxes and sitting-up rooms with experienced staff in attendance all the time, both day and night.

Method (ii) calls for a suitably large loose-box (16 ft x 16 ft for a Thoroughbred mare) with at least a 150 watt electric light bulb and infra-red heater, also a suitable observation point for the watcher – preferably out of sight and sound of the mare.

Method (iii) is to be recommended only where alternative loose-box accommodation is inadequate or cannot be kept really clean. Foaling outside is certainly not to be recommended early in the year when the weather can turn cold and wet suddenly. New-born foals do not have a protective layer of grease in their coats for about the first fortnight and could therefore develop pneumonia. Native ponies, used to living outside all the time, are the only exception to this rule. A further drawback to this method is that most mares foal between 11.00 p.m. and 3.0 a.m. when everything is quiet and they imagine that they are alone, so should she get into any trouble while foaling, it would be very difficult to give help to the mare in the pitch dark even if you happened to see her.

Mares close to foaling must be put in safe paddocks on their own. This means no barbed wire, unguarded ditches or ponds. On no account should they be turned out with geldings or other mares without foals at foot, Otherwise one of these animals might try to steal or damage the foal, before the foaling mare can get to her feet to protect her offspring.

Sitting up with over-due mares or mares which run their milk for weeks before foaling, can be a problem if you are a 'one-man band'. However, modern technology has devised two methods which can assist here:

(1) As from 1983, it is now possible to examine mares' milk in a laboratory for changes which take

place in its composition soon before foaling and so eliminate the need to sit up for long periods unnecessarily especially with mares which run their milk before foaling. Only a drop of milk is required for the test.

(2) Foaling alarms are now available. Transmitters are either attached to a breast girth and roller or encorporated into the back strap of a headcollar. Many receivers may be sited up to two miles away from the transmitter.

If the mare is not already insured on an annual basis she should be covered for foaling and thirty days afterwards; if she is insured annually, the premium should be increased to cover the act of foaling. Once the mare has been tested in foal the unborn foal which she is carrying can be insured against abortion, being born dead, or dying within thirty days after birth. Other foals can be insured from twenty-four hours old.

When your mare foals, or sooner if you sell her before foaling, make a photostat copy of the card on page 39 (Fig. 7). Fill in the details and send it to the stud where the sire stands, This is much appreciated especially where unregistered stock are concerned.

5 Foaling

Choose the largest and warmest loose-box to foal your mare in (ideally 16 ft square for animals over 15.0 h.h.). The floor should be of concrete so that it can be kept clean and disinfected easily. The concrete should be roughened to prevent slipping. All holes or gaps in the lining boards which might allow a foal's hoof to slip through and possibly get trapped should be made up. All draughts must be eliminated by placing rolled sacks across the bottom and down the sides of the door on the outside. To help protect the foal from injury during its first struggles to get to its feet, the box should be well bedded down with good clean wheat straw and the straw built up round the sides of the box to a height of two feet including the area round the door. The box must be mucked out every day and, ideally, the bed left up while the mare is out at grass, to allow the floor to dry out. Cleanliness at the time of foaling is very important as dirty conditions can cause joint ill in the foal (see p. 154).

A loose-box which is sited so that its interior can be seen from your house is ideal. You can then leave an electric light on in the box all night and watch the mare through a suitable window without disturbing her and without her knowing she is being watched. Foaling boxes should have 150 watt bulbs, at least. The top door of the loose-box should be left open to give a better view of the mare. Alternatively a box should be chosen with an adjoining saddle room or feed house which can be used as a sitting-up room. A trap door should ideally be built into the wall to enable the person who is sitting up to keep a watch on the mare without needing to go in to her at regular intervals. All large commercial studs have special sitting-up rooms adjoining their foaling boxes and many of these are equipped with closed-circuit TV.

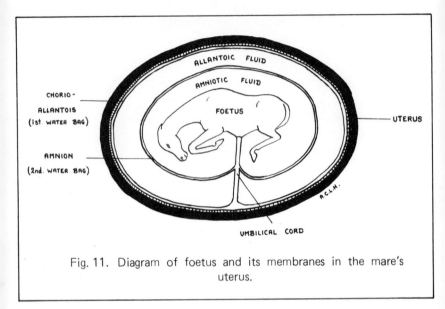

Fig. 11. Diagram of foetus and its membranes in the mare's uterus.

The water bucket in a foaling box should be rubber rather than metal — a metal bucket standing on the floor could cause damage if a foal happened to fall against it.

Hay nets must always be tied up high to eliminate any chance of the foal putting a leg through the net; or better still the hay should be left loose in a corner of the box. When tying up hay nets it should be noted that the net can be tied up far higher if the draw string is passed via the ring through the bottom of the net to the highest point up the side of the net, at which it can comfortably be secured. This is best done by making a slip knot and passing the end of the rope through the loop, so that should the horse pull the loose end of the rope, the knot will not come undone.

As the mare gets nearer to foaling she will tend to get slower and slower and less inclined to exercise herself when turned out. As exercise is essential for an easy foaling it is advisable to walk the mare out for about thirty minutes each day if you have time. When a mare is close to foaling she will tend to detach herself from other horses in her field and wander off on her own.

Just before foaling the mare will tend to get very restless, constantly wandering round her box, possibly getting up and down, swishing her tail and kicking at her belly. Minor birth pains can sometimes be experienced as much as a fortnight or so before foaling, but as the time gets nearer they gradually become more frequent and pronounced.

When the mare seems to be getting near to foaling it is advisable to telephone your veterinary surgeon to ascertain if he is going to be on duty. In this way, if there is an emergency you won't waste time trying to find a veterinary surgeon who is at home. At this stage stitched mares should be cut open by your veterinary surgeon.

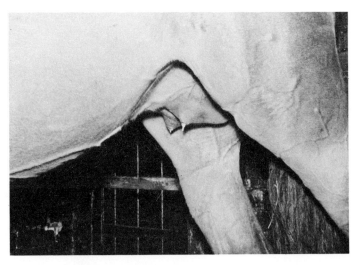

2. When foaling is imminent candles of wax may appear on the ends of each teat.

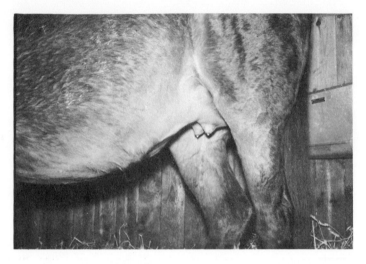

3. In some cases the candles of wax disappear and the mares run
their milk from both teats before foaling.

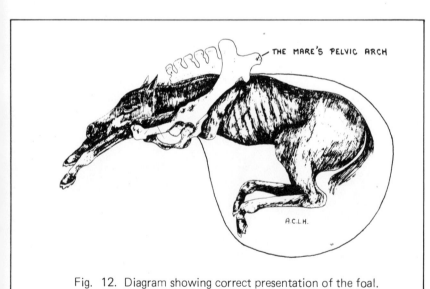

THE MARE'S PELVIC ARCH

A.C.L.H.

Fig. 12. Diagram showing correct presentation of the foal.

The mare's udder gradually gets larger as the pregnancy progresses but will usually go down again after exercise. However, in the final week or so it becomes shiny and remains large even after exercise. Candles of wax may appear on the ends of each teat (plate 2); at this stage the mare could foal any moment. In some cases, the candles of wax disappear and the mare runs milk from both teats before foaling (plate 3).

As parturition approaches, the muscles on either side of the tail gradually become slack and sink (plate 4). At the same time the vulva gradually lengthens and swells.

When the mare starts to foal, she will often walk around her box and paw at the straw. She may sweat profusely. She may get down to foal or commence foaling in the standing position, only going down as the foal begins to arrive – very few mares will foal standing.

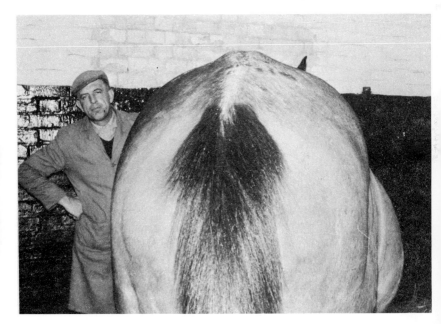

4. The muscles either side of the tail gradually slacken and sink.

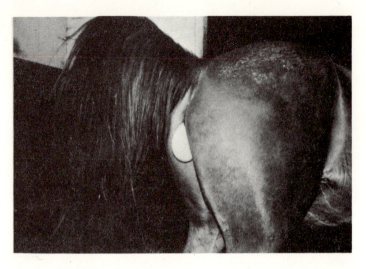

5. The breaking of the water bag is soon followed by the appearance of the amnion.

This first stage of foaling can take from as little as thirty minutes to as long as two hours or more.

The first sign of foaling is the breaking of the water bag (placenta), shown by a flow of yellow/brown water (allantoic fluid) from the vulva. At first only a small quantity escapes and this is not accompanied by the usual 'winking' which normally follows urination but is followed instead almost immediately by a larger quantity of fluid which gushes out of the vulva. This is soon followed by the appearance of the amnion containing amniotic fluid (plate 5).

Soon afterwards the front feet will be seen inside the amnion (plate 6). These should be checked to make sure that one foot is coming in advance of the other — this allows the foal's shoulders to come through the pelvic arch easily. Should an attendant need to pull on the foal's legs, one must always be kept in advance of the other for this reason. (see Fig. 12).

At this stage, the mare may get to her feet only to go

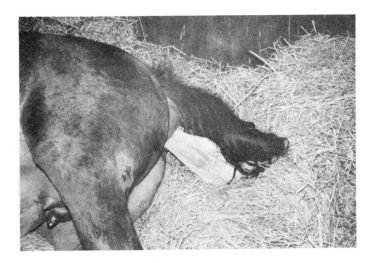

6. Shortly afterwards the foal's front feet will be seen inside the amnion.

down again on her other side. This is thought to be a natural method of correcting a faulty position of the foal.

7. The foal's head will appear between the front legs.

The foal's head will begin to appear between the front legs — in a diving position (plate 7). The amnion will often break automatically as the front feet appear but if not, it should be broken by the attendant as soon as the head is out. If the amnion remains unbroken, the foal will suffocate when the cord severs. The foal itself will often break the amnion by arching its neck and striking with its front feet. The foal's hooves are covered on their edges by a softer pad which protects the mare from injury. These fall off as soon as the foal gets up.

Once the shoulders are out the mare will often rest for some time before completing the delivery (plate 8). When the delivery is complete the mare and foal should be left to rest (plates 9 and 10).

8. Once the shoulders are out, the mare will often rest before completing the delivery.

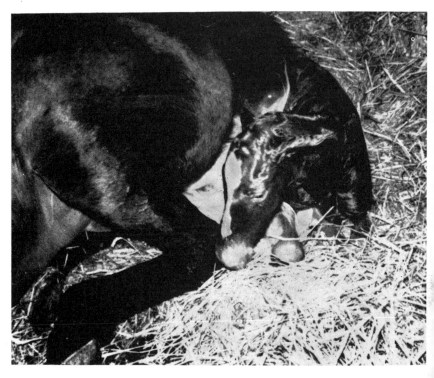

9. Once delivery is complete the mare and foal should be left to rest.

On no account should the cord be broken artificially, as a considerable amount of blood flows through the cord into the foal at this stage. Before birth, the foal receives its entire oxygen supply from the blood flowing through the cord from its mother (plate 10). When the cord breaks, the foal's breathing must be fully established as it then has to breathe entirely on its own. Place your hand under the cord to feel the pulsations of the blood flowing through it into the foal. If the cord breaks before these cease the foal will be deprived of a considerable quantity of blood. On no account should the cord be cut and tied.

10. Do not be tempted to break the umbilical cord until the pulsations have ceased.

11. If the cord is broken or cut artificially, the foal can lose a considerable amount of blood. Sometimes the cord breaks accidentally as has happened here.

12. Gradually the foal will struggle to get up but does not usually succeed in getting to its feet at this stage; if it succeeds the cord will break naturally.

13. When all pulsations have ceased, the foal should be dragged round to the mare's head.

14. By licking the foal the mare learns to recognise her own offspring.

Some authorities like to leave the foal's hind legs inside the vagina as this will help to keep the mare down at least until the pulsations in the cord have ceased. However, others prefer to pull the hind legs clear, without breaking the cord, as this prevents the foal bruising the mare in its attempts to get to its feet.

Under natural conditions the second stage of foaling — from the breaking of the water bag to the breaking of the umbilical cord — usually takes from seven to forty minutes. It should not extend beyond about forty minutes without veterinary aid being sought to make quite sure that nothing is wrong. Often the first and second stages together only take about ninety minutes. If you suspect that there is something wrong and you

15. The after-birth should be tied or knotted up to prevent it flapping round the mare's hocks.

16. The after-birth or placenta: (1) main body of the chorio-allantois (chorion) membranes; (2) pregnant horn; (3) non-pregnant horn; (4) the amnion. Lay out the after-birth as above; check especially that both horns are intact.

telephone your veterinary surgeon, he may advise you to get the mare up onto her feet and walk her round the box until he comes. This will prevent her from straining any more and so exhausting herself.

Eventually the foal will struggle to get up and the cord will break naturally, or, once the pulsations have ceased it may be broken by dragging the foal round to the mare's head (plate 13). Alternatively the cord may be broken by a strong pull with the right hand while the left hand holds the abdomen round the navel. Once it has broken, a packet of sulphonamide powder should be shaken over the stump. This should be repeated again the next day to prevent infection, such as joint ill.

The mare should be encouraged to smell and lick her foal so that she learns to recognise it (plate 14). The mare will probably remain down for at least half an hour after foaling.

The CEM organism has been isolated from the placenta of infected mares. Therefore it is suggested that a swab should be taken from this site which may speed diagnosis.

The after-birth (placenta) gradually separates from the uterus and is expelled usually within an hour of foaling. The after-birth should be tied or knotted up to prevent it flapping round the mare's hocks (plate 15). If the placenta has not been dropped by the morning a veterinary surgeon should be called immediately, otherwise septicaemia may occur. Once expelled the after-birth should be examined carefully to make sure that all the parts are present and no piece has remained inside the mare which could cause septicaemia (see plate 16). After examination it should be taken away and buried. On no account should the owner try to remove the after-birth from the mare himself.

Once the mare and foal get to their feet, the foal will immediately look for the mare's udder (plate 17) and should manage to find the teats within two hours of birth. The attendant should not leave the foal until he has made sure that it has had a drink. Weak foals must be supported to enable them to drink, but other foals are better left to find their own way. Human inter- ference usually delays the time when the foal first sucks.

Occasionally some maiden mares are very ticklish in the region of the udder and need to be held with a front leg up before they will allow the foal to suck; they are usually all right once the foal has sucked and udder pressure is released a little. For this reason, young mares should be well handled between the hind legs before foaling.

It is very important that the foal should receive the mare's colostrum – or first milk – as the foal, unlike the human baby, does not receive any immunity to disease before it is born but obtains its resistance to infection entirely from this foremilk. However, it can only absorb the antibodies for about twenty-four hours after birth, after which time it must start to manufacture its own antibodies. A foal deprived of mare's colostrum is very likely to become ill and may even die. Orphan foals and those born to mares which have run their milk before foaling should be given an alternative supply. (Colostrum will keep indefinitely in a deep-freeze.) Otherwise, your veterinary surgeon should be consulted as he may wish to give antibiotic cover for the first few days, or inject some of the mare's blood, which contains antibodies.

When your mare foals and there is any variation from the normal foaling procedure described, you should notify your veterinary surgeon immediately. If anything is wrong, speed in getting veterinary assistance is impera- tive.

After foaling clean straw should be put down where necessary. Next morning the foaling box must be mucked out as it will be very dirty and wet underneath. This should be done with the door shut, and in sections. Each area mucked out should be bedded down again im- mediately to prevent any chance of the foal slipping on the wet floor.

17. Once the mare and foal get to their feet, the foal will immediately look for the mare's udder.

If the weather is warm and sunny there is absolutely no reason why the mare and foal should not go out into the paddock the day after foaling. If the mare has not been out in the paddock for several days she should be led around for about fifteen minutes before she is allowed to go free, as many mares will kick up their heels on first being let loose and several foals have been injured in this way. Some mares may need to be lunged first without their foals.

From the first day onwards the foal should be handled and taught to lead. Start by pushing the foal gently round the box. To do this, one hand must be placed round the foal's chest and the other round its hindquarters; this way the foal can be led anywhere. Once the foal has the idea of moving around under your guidance, a small, soft headcollar should be put on, and the foal can then be led out with its mother if the weather is fine. A hand should be kept round its hind-quarters all the time to prevent it from running back-wards and throwing itself down. Two people are needed: one to lead the mare, the other to guide the foal.

During this period the foal's feet should be picked up. The sooner this is done the less trouble you will have later. Make sure the foal's headcollar fits it well and has plenty of adjustment to allow for growth.

The mare comes in season again any time from about seven days after foaling. Should you wish to get her in foal again the next year, it is as well to send her away to the stallion about five or six days after foaling, to give her time to settle down before service. Check beforehand if the stud requires any negative genital swabs (e.g for CEM, Klebsiella and Pseudomonas) before they accept your mare onto their premises.

It is not absolutely essential that a mare should return to the stallion for her foaling heat (unless she foals very late in the season), as contrary to common belief this is the time of lowest fertility due to the mare in some cases not having cleaned up properly after foaling – statistics have shown that fertility at this time is as low as 40 per cent of the total number of mares covered as compared with about 67 per cent for later services.

Occasionally some mares exhibit only external signs of oestrus during their foaling heat and do not show signs of being in season again, as long as they are suckling their foals. In these cases the mare should be covered at the foaling heat and examined for pregnancy as early as possible. Should the mare prove to be barren, she should be examined by a veterinary surgeon at least twice a week until she shows internal signs of oestrus. As soon as there is a ripe egg present, the mare should be covered. Some mares do not show any signs of oestrus even at their foaling heat, in which case the mare should be examined by a veterinary surgeon from seven days on-wards after foaling and covered by the stallion the

CODE NO. OF MARE	DAYS ON WHICH MARE WAS SERVED AFTER FOALING	IN FOAL OR RETURNED	RESULT OF SUBSEQUENT SERVICE
1	9 days	In foal	
2	7, 8 days	RETURNED	In foal
3	10 days	RETURNED	In foal
4	8, 9, 10 days	RETURNED	In foal
5	10, 12 days	In foal	
6	9 days	RETURNED	In foal
7	10 days	In foal	
8	10 days	RETURNED	BARREN
9	9 days	RETURNED	In foal
10	10, 11 days	In foal	
11	10, 12 days	RETURNED	BARREN
12	11, 14 days	RETURNED	BARREN
13	8 days	RETURNED	In foal
14	9, 11 days	In foal	
15	9 days	In foal	
16	8, 10 days'	RETURNED	BARREN
17	10 days	RETURNED	In foal
18	9, 11 days	In foal	
19	9, 10 days	RETURNED	In foal
20	9, 11 days	In foal	
21	8 days	RETURNED	In foal
22	10 days	In foal	
23	8 days	RETURNED	In foal
24	8, 9 days	In foal	
25	9 days	RETURNED	BARREN
26	9 days	RETURNED	In foal
27	11 days	In foal	
		40.7% in foal	68.75% in foal

Fig. 13. Random sample of twenty-seven foaling mares.

moment a ripe egg is present. Very good results have been obtained from services under these conditions.

Mares should only be covered at the foaling heat if they are absolutely clean and are not bruised internally.

When the mare comes in season for the first time

after foaling, you will probably notice that the foal's dung becomes loose. This is a perfectly natural occurrence and absolutely nothing to worry about. In order to prevent the foal's coat from dropping out in the region of the dock, any wet dung should be washed off and Vaseline smeared over the entire area in contact with the dung.

If the mare has not been injected with tetanus toxoid about three weeks before foaling the foal should be injected with tetanus serum as soon after birth as possible and again a month later. This will probably happen while the foal is away at stud and will give it protection until it is old enough to be injected with the toxoid. In case the stud do not inject their foals as a routine practice, it is as well to notify them that you wish to have your foal done — this is particularly important in some areas of the British Isles where tetanus is prevalent.

For the first few days after birth the foal should be checked to make sure it is passing its droppings easily. At the time of birth the foal has a hard black substance in its rectum, known as meconium. These are the faeces which have accumulated during gestation and which must be passed before normal functioning of the bowel can commence. The mare's colostrum is a natural purgative but should the foal be seen to strain or have any symptoms of colic a tablespoonful of liquid paraffin or castor oil should be administered without delay. If the foal appears to be in pain it is probably wiser to summon a veterinary surgeon immediately rather than let the foal get any worse. (See p. 155.)

6 The Orphan Foal

Strictly speaking an orphan foal is one which has lost its mother, but other foals may be deprived of their mother's milk due to a variety of reasons, e.g. a mare may reject her foal at birth, or foal down without any milk or become ill at some stage during the lactation period and dry up. Fortunately, all these cases are very rare. However, should anything like this happen you will have two courses open to you: either you must rear the foal by hand or find a foster mother.

1 *Rearing a foal by hand*
The most important thing to remember is that the foal

18. A mare will sometimes accept an orphan foal and rear it with her own, but supplementary feeding of the foals is very necessary.

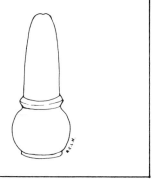

Fig. 14. Teat for feeding foals
(4 in. x 1 in.).

must have colostrum within *the first twenty-four hours* of birth, so that it can develop resistance to most of the diseases with which it will come in contact. Many large equine veterinary practices and studs keep a supply of mare's colostrum deep frozen for emergency use, but where it is impossible to get hold of any colostrum *it is essential* that you should call in your veterinary surgeon so that he can give the foal some substitute colostrum and antibiotic or serum injections. Colostrum from donkeys can be given in the absence of mare's colostrum, but cow's or goat's colostrum is not recommended.

All foals must have some form of milk in order to survive. For the first six weeks the foal must be fed every two hours both day and night; then the length of time between feeds can be increased gradually. Foals cannot be treated like calves, but must be fed little and often; over-feeding at any time will make a foal ill.

At first, the foal should be fed from a bottle fitted with a rubber 'lamb teat', obtainable from most agricultural chemists and measuring 4 in. x 1 in. (see Fig. 14). Milk is also an ideal food for bacteria (germs), so any bottles, teats and mixing utensils used for feeding must be kept absolutely clean and sterilised between each feed. Sterilisation can be carried out with boiling water or by using a dairy or nursery chemical steriliser, such as hypochlorite, in which case care should be taken to make sure that the surface is absolutely clean before treatment. If the equipment is not sterilised regularly the foal will probably start to scour and become quite ill. To avoid digestive upsets the milk should always be fed at the same temperature, preferably blood heat. For this reason, it is better if the same person feeds the foal all the time. Unused milk should be thrown away and never re-warmed.

A satisfactory substitute for mare's milk can be made up easily in the following way:

2 tablespoonfuls of lime-water *not lime-juice cordial*
(obtainable from any chemist)
4 tablespoonfuls of glucose
made up to 1 pint with warm cow's milk.

Goat's milk, when it can be obtained, will give better results than cow's milk and should be fed as a direct substitute for cow's milk in the above recipe.

Where cow's milk is difficult to obtain in sufficient quantity the foal can be given Ostermilk Number 1 at the following rate (which has been formulated for foals weighing approx. 80 lb (36 kg) and over at birth):

From birth to approx. 1 month, per feed
4 measures of Ostermilk
2 measures of glucose
to 8 fl. oz. of water
From approx. 1 month to weaning, per feed
10 measures of Ostermilk
4 measures of glucose
1 pint of water

The Ostermilk should be mixed to a smooth paste with a little cold water, stirring until there are no lumps left, then hot water added and the temperature adjusted to blood heat. Too much liquid will cause the foal to develop a 'pot belly'. The feed should measure between 15–25 fl. oz. or more according to the size of foal.

Foals weighing less than 80 lb (36 kg) at birth should be given the same mixture but less quantity.

Ostermilk, which is really designed for human babies, is a rather expensive method of rearing foals by hand through to weaning. Specialist milk powders are available. They contain all the nutrients necessary for a young foal, in a high-quality milk-powder composition, and should be fed according to the manufacturers' instructions.

Most foals are born with a well-developed suck reflex and those which are slow in this respect will usually develop the desire to suck within two hours of getting to their feet. Ideally the foal should have its first feed within two hours of birth and certainly not later than four hours. Mix up the milk as described above and put it in the feeding bottle. If you have no milk or equipment, it is better to wait until the shops open in the morning than risk upsetting the foal with a home-made substitute. Be careful not to over-feed. When a mare dies foaling, some colostrum can usually be drawn from her udder just before she dies.

The hole in the rubber teat must be large enough to allow a good flow of milk. Give the foal a taste of the milk by squirting a little into its mouth; this should awaken its suck reflex. Making sure your hands are clean, run a little milk onto your fingers and get the foal to suck them, gradually substituting the teat for your fingers. Unless the foal is lying down, it is better to get the foal's hindquarters into a corner, as this will give you more control. If the foal does not start sucking soon, again squeeze some milk out of the teat into its mouth, until it finally gets the idea. When the foal is hungry it will soon start sucking if you persevere.

Try to get the foal to eat some form of solid food as soon as possible. Most will nibble a little warm bran mash from your fingers the day after they are born; something warm and wet will encourage them more than a dry feed. A little bran, boiled linseed, rolled oats and flaked maize dampened with some milk is highly suitable. Once the foal begins to eat, it should be introduced to a 16 per cent crude protein (on a fresh weight basis) dry feed, either nuts or a home-made mix. As the foal eats more hay the protein value of the mix should be increased to compensate for the protein deficiency of the hay. If the foal has access to spring grass no adjustment will be needed.

The ad lib feeding of nuts is possibly preferred to regular feeds, as the foal can then help itself whenever it wishes. A boiled barley and linseed mash containing a raw egg may be fed every night to give variety to the diet. The foal must also receive ad lib hay and water. Care should be taken that foals, and twins in particular, do not get too fat otherwise they may develop epiphysitis.

A foal which is being reared by hand should have some form of companion, otherwise it will tend to become very lonely and will often not do well. A quiet sheep, goat or big dog is suitable. Make sure the foal is kept warm; in cold weather an infra-red lamp may be necessary, or a rug made from a woolly jumper, putting the foal's front legs through the arms.

2 *Rearing the foal on a foster mother*

Most hand-reared foals become very 'humanised' and seldom have much respect for their owners. It is therefore desirable to bring an orphan foal up on a foster mare if at all possible.

If you intend to find a foster mother for your orphan foal, do not teach the foal to drink out of a bowl as it is then more difficult to persuade it to suckle from a mare at a later date. If you do wish to teach the foal to drink out of a bowl or bucket, which saves time, the easiest way to do this is to get it to suck your fingers and then gradually lower your hand into the milk in exactly the same way as one would with a calf. A little food scattered in the bottom of the bowl when the foal has nearly finishing drinking will encourage it to eat.

Rearing a foal by hand is a tedious but rewarding job, necessitating getting up every two hours throughout the night, at least for the first month. The problem of finding a suitable foster mother, without any help, is usually insurmountable but there is an organisation in England known as the National Foaling Bank, which is geared to help anyone who has lost either a mare or a foal. The address and telephone number can be found on page 183.

A suitable foster mother is a mare which is in good health, has suckled a foal before and is the right size. Small pony mares are unsuitable for large Thoroughbred or hunter foals. The mare must have plenty of milk.

Having located a suitable foster mare, the problem is to persuade her to accept the orphan foal. Some mares have a well-developed maternal instinct and will accept any foal readily; others are not quite so keen. In either case one must take certain steps to make the mare believe that the foal she is being asked to foster is in fact her own. Initially, mares recognise their own foal by its smell, that is, the smell of the birth fluids. Therefore this smell must somehow be transferred to the orphan, which can be done in one of two ways:

(1) If the after-birth is still available and fresh, and the dead foal was not diseased, the orphan foal should be rubbed all over with it, making sure that you include its head, legs and tummy. Spread the membranes out over its back as you would a rug. Take the foal into the mare's box, but keep hold of the foal and make sure someone holds the mare. Allow the mare enough freedom of her rope so that she can examine and smell the foal but watch the expression on her face carefully and be prepared to move the foal out of harm's way very quickly should she show signs of anger. If the mare seems happy with her new baby, let the foal go to the udder for a drink. If the mare is still happy, remove the after-birth and leave the pair alone, but watch them carefully through a window for the next few hours. In case of difficulty consult your veterinary surgeon or the National Foaling Bank.

(2) Another method is to skin the mare's own dead foal. Put the skin over the back of the orphan foal and tie it in position with string. However, a word of warning here: you must make absolutely certain that the dead foal has not died from any known contagious disease, such as scouring, because the healthy, orphan foal might very well contract the same disorder. Leave the 'rug' in place for twenty-four hours, then remove it completely. Do not leave it on the foal so long that it starts to go bad — this does not take long under infra-red lamps.

As a safety precaution (particularly in difficult cases) when introducing a mare and orphan foal, the loose-box can be converted into a special 'adoption box', as shown in Fig. 15. The bar must be padded and fixed firmly to the wall either side at flank height. Pins similar to those used in horse trailers are ideal for anchoring the bar as it must be easy to remove.

The mare is jammed between the wall and the bar with her head tied by two ropes into the corner. There

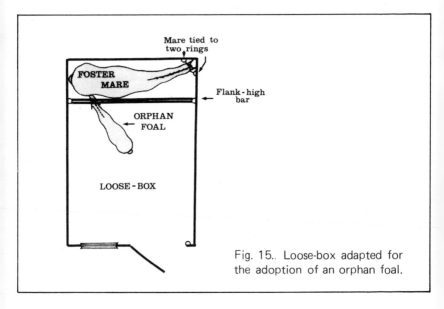

Fig. 15. Loose-box adapted for the adoption of an orphan foal.

must be no room for the foal to walk behind the mare. Watch the mare through a window rather than stay with her; many mares will accept a foal while humans are present, but reject it when left alone. As soon as the mare appears to be happy with the foal and lets it suck, she can be allowed her freedom.

This method works extremely well. It is the reverse of that normally used in the British Isles, where the foal is confined behind a barrier in a corner of the loose-box and the mare is free.

7 Feeding Brood Mares and Youngstock

The ultimate objective when breeding horses is to produce the very best possible foal from the stallion and mare available.

The two most important periods in any young horse's life are, in order of priority:

 (i) the eleven months spent inside the mare (*in utero*);

(ii) the first winter, which is probably after weaning.

The period spent *in utero* is by far the more important as this is the time which determines the size, make-up and shape of the foal at birth. Malnutrition of the foetus for any reason cannot entirely be made good after birth. Therefore, in order to obtain the best possible foal from your mare, she must be fed well throughout the whole period of gestation. As the nutritional value of the grass declines in the autumn, the deficit should be compensated for by feeding a corn ration each day. The earlier a mare is due to foal, the more important this becomes.

Feeding is best done from wooden field troughs which stand on legs and cannot be tipped over by horses pawing at them while eating, or portable metal mangers hung on the fence. One trough must be allowed per animal, and the troughs placed far enough apart to prevent kicking.

When several animals are turned out together in a field one horse will become the boss and the others will assume dominance over each other in descending order. Until social organisation has taken place, kicking and biting may occur. Once the horses have settled down they will tend to thrive rather better than they did before. Size has nothing to do with dominance; the smallest horse quite often becomes the boss.

Where horses are concerned it is imperative that they should be treated as a collection of individuals and never thought of as a group. For this reason, ideally, each horse should be fed separately in its own loose-box. This enables the individual to eat without free competition from its fellows. Where animals must be fed in a large group, for instance when they are turned out together in a field, the food should be placed in troughs far enough away, one from another, to prevent domination of the weaker members of the group. If hay is to be fed, this too should be placed in piles far enough apart to prevent fighting, with at least one more pile than the number of horses present.

Horses should not be fed on the ground, half the food is scattered, pawed into the ground and lost. Also, possibly more important, unless the pastures are worm-free, large numbers of red-worm larvae can be eaten by horses when they crop the grass short in search of the food. The life cycle of the red worm is shown in Fig. 16.

Grass and hay are the keys to your feeding programme. Since nutritional deficiencies in either must be

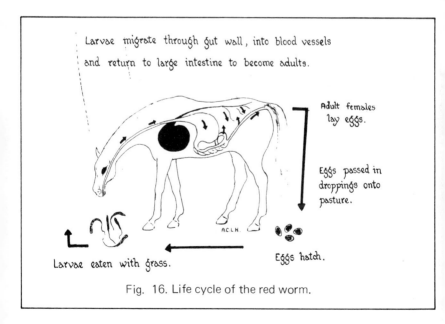

Fig. 16. Life cycle of the red worm.

made good with expensive concentrates, it pays to maintain your paddocks in good condition and only buy top quality hay.

Each class of horse has a different food requirement.

Before she is covered and for the first two-thirds of her gestation period the mare does not need any extra food — her protein requirement is 8.5 per cent crude protein. Therefore she can be maintained on good grass or hay alone, providing they *are* good. If the mare is being ridden she will need more dietary energy than the hay can supply, so must receive oats and bran or horse nuts, according to the work she is doing.

During the last ninety days of her gestation, her foal grows very fast. She requires a higher quality ration, for instance her protein requirement increases to 11 per cent crude protein. If the mare has access to good spring grass for at least twelve hours per day, this should cover her increasing demand for quality food, otherwise some extracted soyabean meal should be added to the oat ration. Do not over-feed the mare as this can lead to a difficult foaling.

As soon as the mare foals and starts giving milk her requirement for all nutrients increases enormously — in relation to her milk yield. Her nutrient requirement then is normally double her requirement during the first two-thirds of pregnancy. Peak lactation occurs between two to three months; after this the milk yield slowly declines and with it the demand for a high quality ration.

A suitable mix for in-foal and lactating mares might consist of:

> 85% crushed oats
> 5% bran
> 10% extracted soyabean meal + vitamins and minerals

Hay usually contains less than 12 per cent crude protein. Barn-dried hay is normally the best and varies between 6–13 per cent crude protein; whereas field-cured hay varies from as low as 2 per cent crude protein to as high as 12 per cent crude protein for a leafy clover mix.

Eighty per cent of the nutrients in hay are contained in the leaf. An estimate of the quality of a bale can be made by examining its colour, smell and amount of leaf. A good sample should appear green and smell sweet with no signs of mould.

In-foal mares, especially those due to foal early, should not be left out at night when the weather becomes cold in the late autumn. They should, however, be turned out or exercised every day up to foaling.

Once the mares are housed at night they should have a feed of oats or stud nuts in the morning before being turned out for as long as possible. On coming in at night they should have a large, dry feed or a warm mash which could contain boiled barley and linseed; the mash should also include any vitamin or mineral supplements and worming powders. Ad lib hay and water should be provided for mares and youngstock and there should be a salt lick in each box.

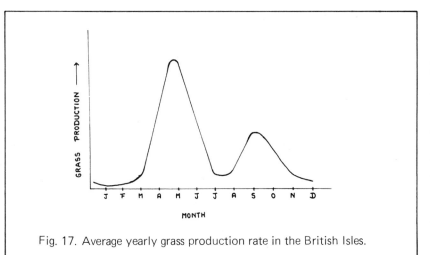

Fig. 17. Average yearly grass production rate in the British Isles.

Brood mares and youngstock have larger appetites than other horses. They will consume 3.0 per cent as compared with the normal 2.5 per cent of their body weight in food per day. However, when feeding horses it must be borne in mind that no two individuals can be treated exactly alike: the good feeder studies the requirements of each horse and acts accordingly; although a basic ration can be devised as a guide to the daily ration this must be modified to suit the individual at all stages. For example, horses which tend to run to fat or get laminitis should have less fattening and heating foods than those which tend to remain thin.

The most common fattening (high energy) foods for horses are: oats, flaked maize, barley and sugar beet pulp. High-fibre foods such as sugar beet pulp should not be given to foals under seven months old, as until this age their digestive system cannot deal with large quantities of fibre. Barley should always be cooked before feeding. When the whole, uncrushed grain is boiled with linseed and mixed with bran, oats and a protein supplement it forms a highly nutritive and palatable feed. The barley and linseed should be soaked in cold water overnight and brought to a rolling boil in the morning. Boiling destroys the poisonous substances in the linseed. Place the covered pan in a slow oven for about three hours, until the grain opens. Allow one

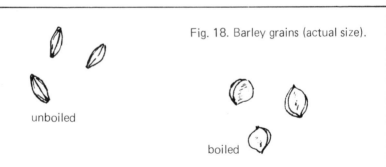

Fig. 18. Barley grains (actual size).

unboiled

boiled

single handful of linseed and about three single handfuls of barley (before cooking) per horse.

To check if you are cooking the barley enough examine your horse's droppings and if these contain many whole grains the mash should be boiled for a little longer.

The most common protein supplements used in horse rations to rectify the deficiency of oats, barley and bran are:

solvent extracted or full fat soyabean meal
dried skimmed milk powder
high grade white fish meal
(Linseed will *not* correct the deficiency.)

When the mare gets very close to foaling she will probably eat less food due to increasing pressure from the foal, which grows very fast during the last ninety days and thus requires correspondingly more nutrients. Consequently, the quality of the ration should be stepped up: less bran should be given, more oats and a protein supplement included; and only the very best hay should be fed. Good feeding before foaling tends to produce larger, stronger foals but excessive feeding to produce above-average foals might lead to a difficult foaling, and is therefore not recommended. A high level of nutrition after foaling enables the mare to achieve the milk yield she is capable of producing and also helps prevent excessive weight loss at this time.

Immediately after foaling the mare should be offered a warm bran mash made in the usual way: to a bucket of bran add some salt and enough boiling water to dampen the bran to a crumbly consistency without making it sloppy; cover the bucket with a sack and allow to cool before feeding.

Young foals should be encouraged to eat from two days onwards. Some warm damp mash will encourage them more than a dry feed. A separate plastic bowl and later a separate manger should be provided for the foal, as most mares, however possessive at other times, will

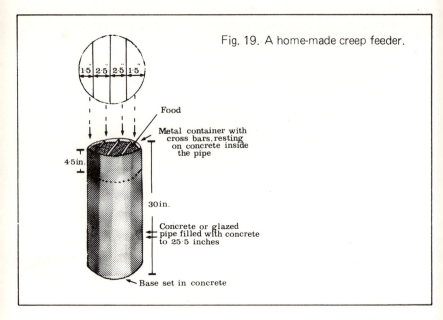

Fig. 19. A home-made creep feeder.

1·5 2·5 2·5 1·5

Food

Metal container with
cross bars, resting
on concrete inside
the pipe

4·5 in.

30 in.

Concrete or glazed
pipe filled with concrete
to 25·5 inches

Base set in concrete

often chase their foals away from the manger at feeding time. Difficult mares should be tied to a ring at the manger until the foal has finished eating. It is usual to supplement the foal's diet with an ad lib supply of 18 per cent protein nuts fed in a creep feeder — obtainable from Horse Requisites, Newmarket, Suffolk; or home-made according to Fig. 19.

The mare and foal should be turned out for longer periods during the day in preparation for staying out day and night once the weather becomes warm enough, probably from the end of May onwards. They can then be left out in their paddock until weaning time.

Except for small ponies, which are capable of living off grass without extra food and indeed would probably get laminitis if they had a corn feed each day, all mares and foals should be fed every day while at grass. In order that the foals can be handled regularly, mares and foals should be caught up each day and brought into their loose-boxes for a feed. Otherwise feed troughs should be used in the paddock.

After weaning the mare should be turned straight out onto the barest paddock available and should not receive any supplementary ration until her milk supply has dried up completely. Unless paddock weaning is being practised the foal should be kept in a loose-box for about one to two days before it can be let out again and then it should come in every night for safety. This can be continued right throughout the winter providing the weather is suitable.

Before weaning can take place the following conditions must hold good:

(i) The foal should be six months old; up to this age its digestive system is not fully developed to deal with a normal high fibre diet.

(ii) The foal must be used to eating the ration it will receive after weaning. On no account must this be changed for about a month after weaning unless the foal develops a disease condition which would necessitate altering the diet.

At weaning time the accent should be on a high quality protein ration, i.e. about 16 per cent crude protein. This can be achieved by adding a protein supplement to the feed, such as dried skimmed milk powder or the specially prepared milk powder for foals which is now available.

For example: 80% crushed oats

5% dried lucerne (or dried grass)

5% dried skimmed milk powder

10% extracted or full fat soyabean meal + vitamins and minerals

This diet should be continued right through the winter period until turning-out time in the spring. By the time a foal is a year old its protein requirement has dropped to 13 per cent crude protein, but it still needs a little protein supplement included in its diet.

Once the spring grass starts to grow the concentrate ration should be cut right back.

If a yearling is allowed to become fat and heavy topped, especially when the ground is hard, concussion occurs at the growth points immediately above the knees. These become swollen, inflamed and sore, causing lameness. This condition is known as epiphysitis. Splints may also develop.

Epiphysitis can occur, for the same reason, in the growth points above and below the fetlock joints of foals up to nine months old. The cure should be obvious — reduce the diet to hay and water only if necessary, and place the animal on a deep peat-moss bed.

During their second winter all youngstock except ponies should come in at the usual time in the autumn; native pony youngstock can lie out if they are provided with some form of shelter and adequate food. It should, however, be realised that any animal living out in all weather conditions requires extra feed to keep itself warm, over and above that necessary to maintain its condition and growth rate.

During the second and subsequent winters a lower protein diet can be fed, i.e. 10–11 per cent crude protein, until the horse has stopped growing. It should then be fed in the usual way according to the amount of work it is doing.

After the first winter, use can be made of cheaper and more fibrous foods, such as sugar beet pulp and carrots. Dried lucerne meal is ideal for brood mares and youngstock, it gives very good results and is high in calcium.

Most of the foods we give our horses contain higher levels of phosphorus than calcium; these include: oats, bran, flaked maize and barley. All horses, but brood mares and youngstock in particular, need higher levels of dietary calcium than phosphorus. A calcium supplement must, therefore, be added to the diet. Ground limestone is normally used except for animals of high economic value for whom the more expensive calcium lactate is often preferred. Two ounces of supplement is usually sufficient to balance a normal ration.

Conversely, horses reared in countries such as America and South Africa, where the diet is based on lucerne rather than grass, may be receiving too much calcium and too little phosphorus. In this case, a phosphorus supplement must be fed.

Vitamin D is essential for the normal absorption of calcium and phosphorus from the food. This vitamin is found in sun-dried plants and dead leaves, hence also in field-cured hay. The amount present depends largely on the length of time the cut grass has been exposed to the sun. It is also formed when the horse is exposed to ultraviolet light. Since this does not pass through glass, some stabled horses could be deficient. Vitamin D is stored in the body but most horses are probably deficient towards the end of the winter. Most vitamin D supplements also contain Vitamin A, the precursor for which is also found in flaked maize, carrots, grass and other green crops. So during the summer the grazing horse should not be deficient in either vitamin. To be on the safe side at other times of the year, diets should be fortified with these vitamins. Other vitamins and minerals should be present in a good diet at satisfactory levels but salt licks should always be available.

8 Weaning

Weaning is the process of separating the mare and foal when the foal is old enough to thrive on a solid diet.

The very earliest date at which this can take place is when the foal is four months old. However, the optimum time for the average mare and foal to be weaned is at six months. There is no definite time at which weaning should occur and it will depend largely on the general condition and health of the mare, i.e. pregnant or barren, and the size and maturity of the foal.

For example, if the mare is pregnant again, even though she may be fat, it could be in the interest of the foetus she is carrying to wean her present foal by six months. In the case of an in-foal mare in poor condition, it might be advisable to wean the foal as early as possible, i.e. at four months unless it too is in poor condition. This type of mare would find it very difficult to maintain herself and her unborn foal as well as support her suckling foal. If a mare loses weight, the energy value of her ration should be increased otherwise her milk yield will fall. Fat barren mares can be left with their foals until the following spring should the owner so desire.

If the foal at foot is in any way weak or ill, it must not be weaned until it has recovered, as the act of weaning is always a shock to the foal's system, and this would undoubtedly set the foal back further. If a sick foal goes off suck the mare must be milked right out at least six times a day to keep her in milk until the foal has recovered.

The actual process of weaning is not difficult and requires only a little planning. The weaned foal should be put in a roomy loose-box and the mare turned straight out into a bare but well-fenced paddock. The paddock chosen must be out of earshot of the weaned foal and should be reasonably bare to encourage the mare to dry off. All fences should be double checked for gaps and weak places before the mare is turned loose.

If possible the foal's loose-box must be large enough to allow it to exercise itself. The water bucket should be placed off the ground, as the foal will probably rush round the box at first and knock over any buckets which happen to be standing on the floor. For the same reason the hay should be placed off the ground, in a manger. Hay racks and hay nets should never be used, since some foals literally climb up the walls. Plenty of bedding should be put down to prevent injury.

As weaning usually takes place in the late summer or early autumn when the weather may be hot, a wire cage should be put over the top door so that this can be left open all the time with no danger of the occupant jumping out. If you haven't got a cage for your loose-box, one can be made quite simply by constructing a wooden frame the same size as the top door of the loose-box, and nailing some half-inch wire netting or weld-mesh over the frame. This can then be tied in place with string to the hinges and fittings of the top door.

If the foal is not already receiving a daily corn ration, one must be given for about two to four weeks before weaning, to accustom the foal to eating corn before the mare is taken away. If this is not done the foal will receive a very severe setback. On no account should weaning take place until the foal is eating corn readily.

On the selected day for weaning, bed the loose-box down well, put in a little hay, a bucket of water and a mash or dry feed according to what the foal has been used to eating. Bring the mare and foal in from the paddock and take them into the prepared loose-box. Put a bridle on the mare with a long leading rein or lungeing rein attached. With the help of a friend walk the mare out of the box, leaving the foal inside, and shut *both* the top and bottom doors. Ask your helper to walk behind the mare with a stick to prevent her from stopping or trying to run back to her foal as you take her to the paddock.

If at all possible have a quiet horse ready to turn out with the mare for company. Loose the mare and her companion and stand by the paddock gate until the mare has settled down completely, which may take several hours. Once the mare is settled and grazing peacefully it is normally all right to return home and see to the weaned foal.

Some people advocate weaning two foals at the same time and putting them together in the same box. This method, as I see it, has one advantage and two disadvantages. The two foals are companions for one another and consequently may tend to settle down quicker and miss their mothers less. Against this they will require a further period of weaning later on when they have to be separated. The other disadvantage is that one foal is bound to become the boss and, unless all the food is provided ad lib or a constant guard mounted at feeding time each day, one foal is going to get less to eat than the other. The author has tried ad lib feeding of stud cubes to foals and yearlings from weaning time onwards with remarkably good results —these youngsters did not ever over-eat but helped themselves whenever they felt like some food, i.e. little and often. The manger was completely emptied at least twice a week to prevent any stale food accumulating in the bottom. This is probably a good method to adopt if you have to go out to work every day during the week and cannot return home at lunch-time to give a mid-day feed to stabled youngstock.

When you return to your foal, after your mare has settled in her paddock, if it is a hot day put the wire cage in place and open the top door of its loose-box.

When the foal is mucked out the door *must* be kept closed and the dirty straw piled against it, then it can be opened just wide enough to allow for the removal of the manure. The top door or wire cage must be kept closed all the time otherwise the foal will quite probably jump out while you are at the back of the box. After the first few days, weaned foals usually remain quietly in their boxes until they are turned out each morning. They may become excited about this but high spirits are something to be expected in young horses.

About two to three days after weaning, foals may be turned out for exercise into a well-fenced paddock. They must be brought in at night for at least two to three weeks. If you decide to turn them out again all the time a dry warm night must be chosen for the first night out, otherwise being so young they might get a chill. Personally, I never leave my weaned foals out at night until they are turned out in the spring as yearlings.

If you have only one foal, a suitably quiet companion should be found or the foal should be turned out to exercise itself for short periods only, a watch being kept on it the whole time. Foals are completely without any fear or knowledge of their own capabilities, which can prove lethal, since they will take on fences, hedges and large gates which would stop any self-respecting adult horse.

When choosing companions for brood mares and in particular for youngstock, apart from being quiet the companion must be in good health and free from worms. This latter is particularly important with regard to donkeys. Nearly all donkeys are heavily infected with lungworm – *Dictyocaulus arnfieldi*. This is peculiar in as much as the normal symptoms of lungworm infestation are usually absent in the infected donkey but the infected donkey is capable of transmitting its worms to any horse with which it is turned out. The symptoms of lungworm infestation are usually a chronic cough with no other evidence of illness or disease; as the condition progresses this cough is usually accompanied by symptoms of broken wind, i.e. the double expiratory lift. If however, a secondary infection should occur the horse will probably develop pneumonia.

The life cycle of the lungworm is depicted in Fig. 20. On being swallowed the larvae pass down into the intestines, where they penetrate through the intestinal wall and enter the lymphatic system, via which they reach the right side of the heart. From the right ventricle they become blood-borne and migrate to the lungs; here they break out into the air passages and eventually reach the bronchi.

Larvae passed in the droppings onto the grass

Larvae climb up the grass and are eaten by the donkey

Larvae passed by the donkey are eaten by the horse.

The infected horse passes larvae which can be eaten by any other horse, as well as itself.

Fig. 20. The life cycle of the lungworm.

Lungworm infestation can normally be eradicated by treating animals with Ivermectin.

A donkey's droppings should be checked for the presence of lungworm larvae before it is turned out with horses. To do this, at least enough faeces to fill a half-pound honey jar should be collected and given to your veterinary surgeon who may send them away to a laboratory for testing.

Many people advocate drawing a little milk from the mare's udder for a few days after weaning but from experience it has been found that this practice tends to cause mastitis rather than prevent it. Very much better results are obtained by turning mares out to grass and just leaving them to walk off the excess milk. Nature works on a supply-and-demand basis and will very rapidly cut off the supply if the demand is no longer there. The great thing is to allow plenty of exercise and a reasonably bare pasture.

The mare's udder should be checked each day after weaning for any indication of mastitis. If this should happen the udder will become hot and painful to the touch and appear swollen and hard. If a little milk is drawn off it will be thin and contain clots. The mare's temperature in most cases is higher than normal and she may appear dull and listless. The condition must be treated immediately because if neglected permanent damage to the mammary tissue may occur with subsequent loss of udder function on the affected side.

Where several mares are kept, the Continental or paddock method of weaning may be used. This is now common practice on many large studs in this country and works well.

To operate this method, the mares and foals are better stabled at night before weaning. In order to wean a foal, the mares and foals are led out to the paddocks in the usual way in the morning. When it is his turn, the selected foal is led into the paddock with his companions and held while his mother is taken away to a distant paddock where she is turned out with other horses. As soon as she is out of sight, the foal is loosed and soon joins his friends. Foals weaned in this way seldom, if ever, miss their mothers and settle down to graze and play almost immediately. Nevertheless they must be watched closely for the first day.

Loose-boxes fitted with weld-mesh grilles in the walls, so that stabled horses can see their neighbours, are ideal for weaned foals. They can, however, be a means of spreading infection at other times.

9 Castration

Castration is the removal of parts of the male sex organs, i.e. the testicles, in order to render the animals more docile and easy to handle. It also removes both the primary and secondary male characteristics, rendering the animal safe to ride and turn out with mares; in some cases it also makes the animal a better jumper. The castrated horse is known as a gelding.

The operation must be performed by your veterinary surgeon and cannot take place until both testicles have descended into the scrotum. In most cases the testicles are in the scrotum at birth, but sometimes they do not appear until the colt is eighteen months old.

Occasionally animals with only one visible testicle are met and, more rarely still, those whose testicles never descend into the scrotum. These animals are known as 'rigs' or more correctly as cryptorchids; those with one testicle down being referred to as monorchids.

The castration of rigs may involve a major abdominal operation, as the undescended testicle often lies inside the body cavity. Uncastrated rigs are usually most undesirable animals; some are capable of getting mares in foal and may exhibit many male characteristics, including screaming and striking with their front legs.

Castration is better carried out in the spring once the frosty nights have gone and before flies become prevalent, or in the autumn when the flies have gone and before the first frosts of winter arrive. This way there is less chance of infection and the horse may be turned out day and night so that he can have plenty of exercise. Exercise is of great importance where newly castrated animals are concerned as it permits drainage of the wound and reduces the chance of swelling in the hind legs and scrotum to a minimum. Should these areas start to swell unduly, your veterinary surgeon should be called in as some infection may be present.

There are two methods of castration: either under general anaesthetic with the animal on the ground, or using local anaesthetic with the animal standing. Either way the horse should be starved the night before. After the operation the horse should be offered a warm bran mash and must be kept under very clean conditions. On no account should he be allowed to lie on dirty bedding, otherwise general septicaemia might result. The horse can be turned out day and night if the weather is warm enough and he has been at grass immediately before the operation. As already stated exercise is of paramount importance.

When the operation has been performed in the paddock under a general anaesthetic, somebody should stay with the horse until he comes round and watch him carefully. Some half-conscious horses will walk into trees or through fences.

Where valuable pedigree stock are concerned, it is often better to leave them uncastrated until they have proved their worth, or otherwise, on the racecourse or in the show-ring. Nothing is more infuriating than to own a champion gelding, which would have been worth many thousands of pounds more as a stallion. Therefore, as a rule, all pedigree stock should be left uncastrated until the end of their yearling year, or in the case of race-horses, until at least the end of their two-year-old season. All other stock should be castrated as foals or yearlings.

10 Handling Foals and Youngstock

For the first year of its life the foal grows very fast and soon becomes enormously strong and playful. It is therefore absolutely essential that it should be handled and halter-broken from a day old onwards, if a struggle with a big foal is to be avoided later on.

Time must be found each day to talk to and handle the new arrival until it will approach you with confidence and let you run your hands over it without showing signs of fear. As far as halter breaking is concerned, for the first few days it is sufficient to push the foal round the mare. To do this, place one hand round its hindquarters and the other round its chest; this way you can steer it anywhere.

As soon as possible, a first-size foal headcollar should be put on. These are preferable to foal slips which tend to ride up a foal's neck and can twist round into its eyes. Do make sure that the headcollar fits well as a foal's hooves are so tiny they can easily slip through a loosly fitting strap. If the headcollar is put on and taken off each day the foal will soon get used to being handled around its head. Some people prefer to leave the headcollar on all the time, in which case it should be removed and cleaned at least once a week — a foal's skin is very soft and can easily get chafed by stiff or dirty tack. If the headcollar is removed for cleaning every week adjustments for size can be made. Since foals grow very fast, without regular weekly checks the headcollar can easily rub the foal's face until the skin is raw.

Before you take your mare and foal out of the loosebox, or from the paddock if your foal has been born outside, it should be halter-broken. Most foals will run round the mare in wide circles while she is being led and in this way they can literally 'run into trouble', especially if there are any sharp objects such as harrows or barbed wire lying around.

Foals will often run backwards and throw themselves on to the ground the first time they feel a pull on the headcollar, so a soft landing is essential. Wherever possible, care should be taken to prevent the foal flinging itself down, as an awkward fall could possibly cause a fractured skull or damage to the spinal column, which might result in wobbler symptoms – inco-ordination of the hindquarters – which are incurable.

Slip a leading rein, at least eight feet long, through the headcollar, keep your right hand firmly round the foal's hindquarters to push the foal forward when necessary; your other hand, holding the rein, can be placed round the foal's chest to prevent it from rushing forward. Once the foal is moving forward well, your hand can be transferred to the base of the neck and the foal can then be led in the usual way.

Should the foal fling itself over backwards, allow the rein to go slack immediately and don't apply any tension to it until the foal is on its feet once more. The moment it jumps up it will undoubtedly run to the furthest end of the rein but with a little patience you can get your hand round its hindquarters and start pushing it forward again.

If you ever wish to lift your foal onto its feet straighten its front legs, then place one hand round its hindquarters and your other hand round its chest and lift. Never try to lift a foal by clasping your hands under its tummy, as you could damage its internal organs and ribs.

To catch a nervous foal by the headcollar, move your hand slowly under its muzzle. The foal will drop its head to examine your hand, which enables you to slip your fingers through the back of the noseband.

When leading young horses from the headcollar, never attach the rein but slip it through the noseband. If the horse gets away from you, the rein will fall off and not become wrapped round it legs and frighten it.

When you venture outside the loose-box with your mare and foal, ideally the mare should be led in front

19. Correct method of lifting a foal.

with the foal close to her near side. Should the foal get left behind, the mare must be stopped immediately and the foal allowed to catch up, otherwise the mare will become very upset. Some mares, however, cannot bear to go anywhere without being able to see their foals all

Fig. 21.. Teaching a foal to lead double with its dam.

MARE   FOAL

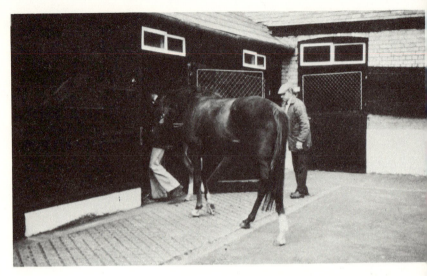

20. Lead the foal into and out of the loose-box in front of the mare, never at the same time.

21. Lead the foal in your right hand and the mare in your left hand. Initially, ask a friend to walk behind the foal and urge it on whenever it stops.

the time, in which case the foal should be led in front of the mare – this is often especially true when the mare is asked to go into or out of her loose-box, trailer or horse-box (see plate 20).

Once the foal is leading well from the nearside, it is a very good plan to practise leading it from the offside too, as this may be useful one day. If you are ever short of help and need to lead the mare and foal one in each hand, it is customary to lead the mare on your left-hand side and the foal on your right (plate 21). However this must be reversed if the mare will only lead from her nearside. To practise this, ask a friend to walk behind the foal and urge it forward whenever it stops.

During the first fortnight of its life it is advisable to start picking up the foal's feet – the sooner it gets used to being handled the less trouble you will have later on when it gets bigger and stronger. Whenever your black-smith comes to trim your mare's feet, ask him to go round your foal's feet too, even if there is nothing for him to do. The foal will then get used to the blacksmith from an early age. It should, however, be noted that the young horse is growing at a fantastic rate during the first two years of its life, so neglected feet at this time, above any other, can lead to permanent twisting and damage to the legs in a very short space of time. Therefore, a young horse's feet must be rasped at least once a month regularly: always remember the old adage, 'no foot, no horse'.

When your foal is still a baby, on no account be tempted to play with it, most especially if it is a colt: if you allow a foal to chew and suck your fingers at this age, don't be too surprised if it bites you later on; if you teach it to shake hands or rear, don't be surprised if it strikes with its forefeet and rears when it gets bigger.

A young foal, like a child, has not matured mentally and this must be taken into consideration when dealing with it. At the same time it is absolutely essential that you take a firm line and do not let it get away with any disobedience. If you do, like a child it will be twice as bad next time. Sometimes, though, it is better to ignore small offences, otherwise you may find yourself constantly scolding your young horse, which is not good for it or for you. For example, most colts tend to nip in play and this can be ignored up to a point. Horse psychology, like child psychology, is something which comes with practice.

If your foal is a big one, i.e. the type of strong Thoroughbred or hunter foal which will make over 16 h.h. when fully grown, it is advisable to mouth it as soon as it is weaned. Otherwise as your foal gets bigger and stronger there will come a day when you will find yourself unable to stop it with a headcollar alone. The young horse should never learn the dangerous habit of setting its neck and charging, as once learnt this is a very difficult trick to eradicate.

In order to mouth your horse and get it accustomed to having a bit in its mouth you will need:
(a) a straight bar mouthing bit with keys; and
(b) a sound bridle fitted throughout with buckles or studs, of the correct size but with a larger browband for easy fitting and no noseband.

To put the bridle on for the first few times, remove the bit, undo the cheek-piece on the nearside of the bridle and slide the browband off, so that it is attached only to the offside of the headpiece. Pass the headpiece over the poll and hold it on the nearside. Pick up the browband, pass it across the front of the forehead and slip it on to the headpiece in the usual way; the throat

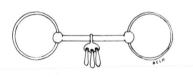

Fig. 22. Mouthing bit.

lash can now be buckled allowing at least the width of three fingers between it and the cheek. It is not advisable to attach the bit to the bridle before it is put on as the young horse will often throw its head round while you are putting on the bridle for the first few times, and in so doing could bruise itself on the side of the face with the loose bit. Attach the bit separately when the bridle is in place. Buckle one ring of the bit to the far side of the bridle in the usual way, then pass the bit across below the horse's teeth. Insert your thumb into the corner of the mouth on the nearside, open the mouth gently and slip the bit in, securing it to the bridle on the nearside. Never try to force the bit between the horse's teeth.

For young horses it is preferable to adjust the bit rather too high in their mouths, i.e. so that it wrinkles the corners of the mouth and prevents them from learning the habit of getting their tongues over the bit, especially during the early stage of mouthing and playing with the bit and keys.

Mouthing must be carried out in a loose-box but first check that there are no nails or projections sticking out on which the horse can catch the rings of his bit. For this reason the top door of the loose-box must be kept shut, otherwise the horse will put its head over the door and could possibly get caught on the door fastenings, which might result in a broken neck.

For the first day the bit should be left in for ten minutes only, so that the horse's mouth does not become sore. For the second day the bit can be left in for half an hour and thereafter for half an hour to an hour twice a day for two weeks. At the end of this time the young horse should be ready to lead out in hand.

When I come to lead a young horse from a bit I always use a straight bar flexible rubber one − the type which is constructed on a chain and can be bent double − not a vulcanite or nylon bit. The soft, flexible rubber bit is, I feel, less likely to damage the horse's mouth should he play up and you have to pull him. Never attempt to lead a horse off the bit until it is completely obedient to your voice, otherwise you will find yourself having to pull on its mouth all the time.

The leading rein, which should be at least twelve feet long, can be attached to the bit in one of two ways:

(a)   by using a three-way bit coupling; or
(b)   by buckling the rein on to the offside ring of the bit and simply passing it across the back of the jaw and through the nearside ring to the leader's hand. In this way the rein rests lightly behind the jaw until the horse requires restraining, when it is pulled tight.

At this stage the first essential for any well-schooled riding horse can be taught, namely: *controlled forward impulsion.* The other two major principles, *correct bend in all movements* and *even rhythm at all paces*, will of necessity come later.

To deal with the first principle – the natural instinct of all horses is to move forward and the young horse must be encouraged to move forward freely in front of you. This will enable you to take up a position by his shoulder with your right hand about a foot or more from the bit and your left hand holding the loop and slack of the rein. Never wrap the rein round either hand, especially when leading from a headcollar. In the event of your horse taking fright and bolting, the rein would tighten on your hand, you would be unable to let go and would be dragged.

In order to create controlled forward impulsion, you have two aids at your disposal:

(a)   your voice;
(b)   your whip.

The latter must be thought of and used solely as an extension of your arm; every horse will do more for the man he loves than for the man he fears.

The whip should be carried in your left hand and used on the flank or quarters at the same time as your voice, to encourage the horse to walk on. It can also form a useful aid when turning the horse away from you, in which case the handle end should be held up to the horse's face to get him to bend his head and neck to the right — never ever hit him on the face.

Some time before you start to lead your horse out in hand and school him, decide exactly which words you intend to use and always stick to them. Make your words of command short and simple so that your horse can learn to understand them quickly. Be ready to make much of him when he does well and always finish a lesson on a good note, i.e. the moment he has understood and obeyed you. This way he will return to his paddock or loose-box in a happy frame of mind and will tend to do better the next day. Never ever repeat an exercise the horse has just done well otherwise he will probably do it badly and you will be 'back to square one'. Horses always remember a lesson better if they are allowed to sleep on it. With young horses in particular, it is essential to keep your lessons short.

As well as being taught to walk and halt in hand, the young horse must also be taught to trot and rein back — the latter should be done by pressure on his chest and not by hauling on his mouth. At most, only one or possibly two steps backwards should be attempted, accompanied by the word 'back'. The horse should immediately be asked to walk on again to prevent him from learning the habit of running backwards.

Once the horse is a year old, he can be introduced to the lunge; this is the next step in his education after being led and follows on in a perfectly natural order from it.

The degree of maturity of the young horse as well as its conformation should be considered before lungeing is started. An immature horse, one in poor condition or with weak hocks, must be left until it is much older and better able to withstand the work, otherwise irrevocable damage may be done to its limbs.

The necessary equipment for lungeing is:
(a)   lungeing cavesson;
(b)   lungeing rein;
(c)   lungeing whip;
(d)   four brushing or polo boots.

The cavesson should be as light and strong as possible with a felt or well-padded noseband. Unlike in an ordinary bridle, the noseband and throat-lash straps are done up as tightly as possible. This is because a loose noseband would chafe the nose and unless the throat lash is tight the cavesson will work round into the horse's eyes.

The lungeing rein is attached to the front ring on the cavesson and should be approximately twenty-five feet long. When buying a lungeing whip the very lightest one possible should be chosen, as this will be less tiring to use.

When a horse is lunged, and in particular a young horse, it is essential that he has some form of protection on his legs to prevent damage should he knock himself. For this reason brushing or polo boots should be worn (those with Velcro fastenings are ideal as they are easier to do up and undo than buckles). Alternatively gamgee tissue or sponge-rubber sheets can be applied to the legs under an ordinary exercise bandage.

From my own experience I have found it advisable not to use side reins when lungeing a young horse as they tend to stiffen the horse and, apart from learning obedience, the object of lungeing is to obtain a correct bend and cultivate free forward movement. Should the horse try to stiffen and turn out of the circle, perhaps towards his loose-box or paddock gate, it is far easier to pull his head round quickly without side-reins than it is with them.

To teach your horse to go on the lunge, put the lungeing tack on as described above and lead him out to the paddock. Using the words of command you have already taught him, lead him round in a circle to the left but gradually let him have a longer rein. If your original lessons were a success, you should find that your young horse will walk round you in a circle on the lunge without any trouble. If you get into difficulties, ask a knowledgeable friend to help you by leading the horse round.

Continue to walk in a small circle yourself until your horse is really balanced – this can take several months, after which time you can stand still in the middle of the circle and let the horse go round you. Grind one heel into the ground and use this as a pivot; change your pivot foot when you change the horse's direction. When lungeing keep level with a point just behind the horse's shoulder and keep your whip level with his hindquarters; this is still used as an extension of your arm to keep the horse moving forward.

It is essential that you should change rein every few minutes when lungeing otherwise the horse will tend to become one-sided if lunged continuously in one direction.

To change rein on the lunge, first halt your horse and walk up to him slowly, taking up the slack of the rein as you go. Talk to the horse as you approach him and then make much of him. Walk round to his other side, slowly transfer the rein into your other hand as you change sides and at the same time change your whip into your other hand by passing it behind you and away from the horse. Tell your horse to walk on, and start by lungeing on a small circle, gradually increasing the size as the horse settles down again.

When the horse first goes out on the lunge in the morning he will probably shoot off, bucking excitedly. In this case you must just stand in the middle of the circle and wait until he has settled down before you ask for any obedience, as in this frame of mind the horse is bound to ignore you for a while.

Any words of command can be used as long as they are clear and intelligible to the horse and always given with the same intonation of voice. The words of command I use for my horses are:

for increases of pace

Walk on; T – rot (with a long T); Can – ter (in two syllables).

for decreases of pace

Walk (long, drawn out); and a long, drawn out Whoa for the halt.

Should a slowing-down word of command ever be ignored, repeat the word and give a sharp jerk on the lungeing rein. The action is transferred through the tightly fitting noseband of the cavesson to the horse's nose and usually slows him down immediately. Therefore a horse must never be lunged directly off the bit as a similar action would damage the bars of the horse's mouth. If an order to increase the pace goes ignored, crack the whip behind the horse but never actually hit him. To prevent a horse from cutting in on the circle, move your lungeing whip across in front of you until it is level with the horse's shoulder.

It is very necessary to teach young horses to go correctly on the lunge, as this is a good means of exercising them when the fields are too wet to turn them out, say, in the winter or early spring; it is also a useful means of exercising show horses.

Further information on lungeing and long-reining can be found in *Young Horse Management*, published by Pelham Books.

11 Preparation for Showing in Hand

If you are going to show your mare and foal or youngster in hand, it is essential that you decide exactly how many shows you intend to enter right at the beginning of the season and stick to this number. Showing can become a very expensive hobby, even if you happen to win prizes most of the time. It is therefore essential that you should have your own transport.

It is better to make up your mind early in the year which shows you intend to enter, especially if you wish to go to some of the bigger ones, as the closing dates for entries are often very early in the year and many months before the actual show date.

If you are not sure which shows to patronise or indeed where to apply for the schedules, the best thing to do is to buy copies of the special show numbers of *Horse and Hound* magazine which come out in two parts at the beginning of March. This will give you a complete list of almost every show to be held during the year, together with the secretaries' names and addresses. It also contains many helpful articles connected with shows and showing as well as a very comprehensive list of stallions at stud in the British Isles.

When you have decided which shows you might like to enter, write to the secretaries concerned and ask them to send you their schedules and the required number of entry forms, i.e. one for each horse. The schedule will give you a list of classes, judges, prize money and special prizes as well as the show's rules and regulations.

Small shows will often accept entries on the actual day of the show for an additional entry fee. This must, however, be checked beforehand in the schedule as not all shows offer this concession and nothing would be more disappointing than to arrive at a show only to find that you cannot enter.

When completing your entry form make certain you fill in all the relevant details and sign the form where required. It is also important that the correct entry fees are included, as without these the entry will not be accepted.

At most of the larger shows you will find that you can hire sleeping accommodation for yourself or your staff. These are sometimes provided with a bed and mattress but at other shows they are quite bare. Remember to pack plenty of padlocks so that you can keep your possessions locked up. It is usually safer to keep all your show equipment in a large box which is fitted with a key, than to leave any equipment unattended, even for a short time, on a show ground. If necessary, have a box specially made for the purpose.

It is a very good idea to keep a notebook of every show attended, the judge, the horse shown and the prize won, if any. In this way you will be able to see at a glance exactly which judge likes which horse or type of horse. This sort of information can be most useful when you make your show entries in future years.

If you have followed the schooling programme suggested in the chapter on handling foals and youngstock there will be no need to carry out any extra schooling sessions especially for the show ring. The only requirement for showing in hand is that the horse should be able to lead well at both the walk and trot. A certain degree of high spirits is usually overlooked as far as foals and yearlings are concerned. In fact it is preferable if they are on their toes in the show ring, as they then look better than animals which are very well behaved but half asleep.

It is an absolute must that any horse which is going to be shown regularly should be good to load into a horse-box or trailer. Otherwise you will find yourself at a show with a reluctant horse, surrounded by dozens of people all offering advice and, as the crowd gets larger, the horse becoming more and more determined not to be loaded. Therefore, if you have your own means of transport, it is a very good plan to practise loading and unloading the horse long before the actual show date. To be on the safe side the earlier you get a foal accustomed to being loaded and travelling the better. While the foal is still with its mother and is quite small, loading will present few problems unless the mare is bad to load. Once the foal is weaned it should be taught to load as soon as possible. To do this I load my foals into a trailer every day, take them up the road to turn them out and bring them back to their loose-boxes in the trailer at night. In this way they get used to being loaded from an early age, and indeed they will usually walk into the trailer on their own after only a week or two. I do, however, appreciate that this method would not be very practicable for someone who has a number of weaned foals to turn out — obviously it would take far too long. But for the small breeder with only one or two foals each year, it is an admirable method of getting them used to travelling.

For older horses which prove difficult to load, a lungeing rein, attached to the side of the trailer or horse-box and passed round the back of the horse just above the hocks, can be slowly pulled up tightly on the far side of the trailer or horse-box to encourage the horse to walk in easily. The horse's head should not, however, be pulled hard, nor should the person leading the horse turn round and look at him. If the horse rears and is in danger of coming over backwards, the lungeing rein should be dropped immediately. A few oats in a bowl will often entice a horse into a horse-box, or failing this, if the horse is being genuinely stubborn and is definitely not frightened, one or two hard bangs on the rump will often do the trick. Make sure the horse is not frightened first, otherwise you will do more harm than good.

If the horse can be trusted not to kick, an alternative method to the lungeing rein is for two people to link hands behind the horse and lift it up the ramp. With a young horse which is used to having its feet picked up, they can be placed on the ramp, one at a time. It is a good idea to scatter straw down the ramp, to make it look more inviting.

The best time to trim your horse in preparation for the show season is as the spring approaches and he starts to lose his winter coat. At this time of the year you will find that the mane and tail hair comes out much easier, as does any surplus hair around the fetlocks and head.

In breeds where a clean heel denotes quality as in Thoroughbreds, Arabs, hacks, hunters and riding ponies, this is best done a little at a time, otherwise if there is a lot of hair, the area will get very sore. On no account should the hair be cut with scissors, as this will show as ridges down the fetlock and look terrible. The fetlock should be plucked out completely.

Any surplus hair on the face should either be plucked out if it is loose or it can be singed off with a lighted taper. The tough whiskers around the nostrils and mouth can be trimmed carefully with a sharp pair of blunt-ended scissors.

If you do not intend to show your mare or youngster with its tail plaited, it should be pulled. This must be done over a long period and, like the fetlocks, only a few hairs taken at a time. A tail bandage should be put on the tail all the time that the horse is in his loose-box to encourage the hairs to lie down.

The mane should also be pulled, preferably up to about three or four inches long, so that it will need pulling only once more before the show season is over. Unlike the tail this can be done all at once. The mane should be well-brushed out so that it is all lying on the offside of the neck. Use a flat-topped metal comb for pulling. Taking only a few hairs at a time, back-comb them to the required length, wrap the longer hairs you are holding in your left hand, round the comb and pull them out. Continue down the mane keeping a perfectly straight line along the neck. If the mane is too thick, it can be thinned by pulling hairs out from the underside but never from the top. A thin mane can be shortened by using a sharp knife instead of pulling the hairs out by the roots after you have back-combed the hair, the hairs should be broken across the length of the blade as this gives a more natural appearance. On no account should a mane ever be cut with a pair of scissors.

All horses' feet should be trimmed regularly – at least once every six weeks – but where you intend to show a horse its feet should not be touched preferably for about two weeks before the show. Otherwise, if the ground happens to become hard, your horse may go foot-sore if he has been newly trimmed. Some people always put light shoes on the front feet of all the young horses and brood mares (but not foals) they show. Minor faults in the front feet can sometimes be rectified by the use of shoes but from my own experience I have always found that more horses move better without shoes, than do with shoes on. There are of course exceptions to this rule; mainly horses with split walls or thin soles.

Horses which turn their toes in or out can have this fault rectified by careful trimming or shoeing from an early age. The outside of the shoe can be built up to throw a toe out, or built up on the inside to bring a toe in. An older horse with pin toes – toes which turn in – can be made to look straighter if the clips on the shoes are offset slightly to the outside, while if a horse has front feet which turn out, clips placed slightly to the inside of front, will tend to make its feet look straighter.

When preparing horses for the show ring, especially if these horses are youngsters out at grass all the time, their coats must be kept in very good condition so that a surface shine can be obtained at a moment's notice without necessitating the removal of any grease from the coat. This is best done by feeding linseed in the ration each day – as described in the chapter on feeding.

Show horses should be fed every day even if they are out at grass day and night. To get them into the required condition for the show ring, about half a two-gallon bucket of mash per horse, per day, is usually sufficient for horses kept at grass all the time, until the quality of the grass declines to such an extent that additional feeding becomes necessary. Small ponies are probably far better without an extra feed, except for hay, when at grass, as they are usually prone to attacks of laminitis, especially if they receive corn.

22 and 23. The author's mare Chunky Clemantine demonstrates the difference a noseband makes to the appearance of a horse's head. The bridle used here is an in-hand show bridle, sometimes known as a stallion bridle.

23.

When horses are shown straight off grass they are inclined to 'dry up' on coming inside, even for one night. It is therefore essential to feed cut grass as well as hay when you bring them in for a show, otherwise in only a matter of hours a horse can appear really hollow flanked and run up.

Very little special equipment is needed for showing in hand. The main requirement is that any bridles, headcollars or leading reins taken to the show should be spotlessly clean and shining. Brood mares can be shown in special in-hand showing bridles or double bridles – the latter are more usual when showing larger ponies, hunter or Thoroughbred brood mares. All show bridles must be fitted with nosebands, otherwise the horse's face will appear bare and unfurnished (see plates 22 and 23).

Double bridles used for showing should preferably be stitched or fitted with studs rather than buckles, which tend to look clumsy. All metal parts of the bridle must be clean and shining — this includes any buckles as well as the bits. Stainless metal buckles are the easiest to keep clean, but rust can be removed from other types of buckle by rubbing them with very fine sandpaper.

For ease of work and first-class appearance there is nothing to beat stainless steel for all bits. Although nickel can look nice, it does require an enormous amount of work to keep it up to show standard, without any mark or stain.

Foals should be shown in brass-mounted headcollars fitted with narrow browbands. These headcollars are usually made from leather but any other serviceable material would be acceptable. Some people like to use white webbing headcollars for very young foals. These look nice but have the disadvantage of having to be cleaned very regularly with white canvas shoe cleaner which might have a disastrous result if it rained. Brass mountings look far better than tin, but all the brass work must be polished and shining. Quiet yearling fillies may also be shown in headcollars only, without bits, but entire colts, geldings and older fillies are all safer with bits in their mouths. For leading young horses in hand, as already explained, I prefer to use a flexible rubber bit and this can be attached quite satisfactorily by means of special leather bit straps to the headcollar, or better still they can be put on to a special brass-mounted in-hand bridle. Either method is equally correct but the in-hand bridle possibly looks a little neater.

I always clean the top side of the leather on my show tack with dark tan shoe polish and apply saddle soap to the reverse side only. After a few applications of polish, the leather begins to adopt a lovely reddish tone, which looks well on most horses.

Extra shine can be obtained by putting polish on in thin layers and rubbing up small areas at a time to a shine with a little water. A glass-like surface can be produced in this manner, which is useful for the front of nosebands, browbands and for the toes and heels of jodhpur boots. Do not be tempted to do this too often, however, as the leather may eventually crack.

As far as possible, oil should not be put on show tack unless it has become very hard. Oil will cause the leather to become dark and make it impossible to obtain a shine on the top surface.

There are two types of leading rein in common use at shows: leather leading reins and white web leading reins.

Leather leading reins should never be cleaned entirely with saddle soap – in the rain the leather will become slippery and if the horse takes a pull at this time the rein will simply slide straight through your hand. These reins are the best and most reliable type, as they are unlikely to rot and break suddenly, as do web leading reins. They should be cleaned in the same way as the show bridles, with saddle soap applied to the reverse side of the leather only, and shoe polish to the skin side. Apply the saddle soap very thinly to the reverse side, otherwise the rein may still become slippery.

White web leading reins are much cheaper to buy than the leather variety but they last for a much shorter time, probably due to the fact that they have to be cleaned with white canvas shoe cleaner every time they are used and this tends to rot them eventually.

The leading rein can be attached to the bit in one of two ways. Firstly it may be buckled to the far ring of the bit and passed through the nearside bit ring to the leader's hand; this method probably gives the greatest control but is consequently harder on the horse. The second method is to use a bit coupling, which is usually made from leather and simply buckles or clips onto the rings of the bit on each side and has a brass ring in the middle of it to which the leading rein is attached; this method produces a more even pressure on both sides of the horse's mouth. The leading rein can also be made with a bit coupling already attached known as a Y-coupling. Make sure that the horse cannot get hold of the bit coupling with his teeth, otherwise it may get caught on his bottom jaw and cause him to rear up and come over backwards. This can be avoided by using a three-way coupling, which also fastens onto the noseband or throatlash. If your horse is very quiet but you still wish to show him with a bit in his mouth, which gives a more finished appearance, the leading rein may be attached to the noseband of the bridle and the horse led from this entirely; or it may be attached to both the noseband and bit coupling to lessen the effect of any pull on the horse's mouth and transfer some of the pull to the noseband.

The appearance of a beautifully turned-out horse can be spoilt by a badly turned-out leader, which after hours of work on the horse is a great pity, to say the least.

For showing in hand you should be able to run fast when the occasion warrants, so it is essential to wear a pair of shoes which are easy and comfortable to run in — long riding boots and high-heeled shoes are unsuitable. Jodhpur boots look smart when worn with fawn trousers, or flat shoes with a skirt.

Men usually wear a neat country-style suit or sports jacket, cavalry twill trousers, a collar and tie and usually a trilby hat, cap or bowler hat. In warm weather a cotton jacket can be worn, but however warm the weather it is simply not done to appear in the show ring in your shirt sleeves.

Ladies can appear in the ring in a skirt but are more commonly seen in a sports jacket and fawn or light-coloured stretch slacks, with a white shirt and tie or jumper. Many people wear hard hats, bowlers or head scarves when showing horses in hand.

The following is a checklist of the articles you will need to take to a show if you intend to stay overnight:

headcollar
rope
hay net
water bucket
feed
hay
feed bucket
scoop
skip
mucking-out fork
roller and tail guard
knee pads
leg bandages and gam-
    gee tissue
tail bandages

lungeing cavesson
lungeing whip
lungeing rein

plaiting equipment
  thread
  blunt needles
  scissors
  box to stand on
wire cage or rails
show tickets, numbers,
    documents, passes

grooming kit
  body brush
  dandy brush
  water brush
  curry comb
  mane comb
  hoof pick
  stable rubbers

show bridle
lead rein
cane
dog chalk block
hoof oil
rag
Vaseline
disinfectant spray
fly repellant spray
shoe polish
saddle soap
cloths and sponges
brass cleaner

collapsible bed
sleeping-bag, pillow
blankets
torch

12 Showing in Hand

When you arrive in the vicinity of the show ground, you will find route signs indicating livestock, horses, etc., which direct you to the correct entry gate. Be prepared to show your tickets to the man standing at the entrance gate and make sure that any livestock passes are signed where necessary. Many of the larger shows have a veterinary surgeon on the gate to check all horses before unloading to make sure that none has a contagious disease, such as a cough or runny nose.

Once your horse has passed the veterinary inspection make your way to the horse lines where you should find the horse foreman. He will tell you exactly where your loose-boxes are and will probably have the ring numbers. At some shows you can drive down between the rows of boxes to unload and load your horses, which saves walking them from the horse-box parking ground.

When you have found your loose-boxes, check that the manger is clean. If it still contains stale food from a previous horse this must first be cleaned out and the manger and walls of the box sprayed with a disinfectant spray in case the previous occupant had any disease. Straw is provided at some shows so double check the schedule before you leave home and bring sufficient straw with you if necessary. If the boxes have a canvas sheet across their front, this should be tied up with string or baler twine before the horse is unloaded and asked to walk into the box, otherwise it will flap around his head even if held up and may frighten him. If you have hired an extra loose-box you should unload all your equipment before moving your horse-box, as this will save a lot of carrying later on. When everything has been unloaded, take your horse-box to the parking ground.

At some shows the loose-boxes have rather insecure door fastenings and it is therefore safer if the bolt is tied with string or padlocked before the horse is left, as the bolts on these doors can easily slip open.

Preferably before you park your horse-box, leave the horse with a full hay net, feed and water. Water taps are usually located at the end of each horse line with a galvanised tank under the tap. Always draw your water directly from the tap; never dip your bucket into the tank, as this water is often stale and contaminated by other people's dirty bucket bottoms.

As soon as you have found somewhere to park your horse-box return to the horse lines and make sure your horse has settled down.

If the show is providing greenstuff for its livestock exhibitors, it is better to go to the forage yard as soon as possible as they often close fairly early in the evening and don't open again until the following morning. Where forage tickets are issued, they are either sent by post with the other show tickets or obtained from the horse foreman. Due to the high price of forage these days, shows no longer provide hay.

Meals are available on most show grounds in the members' and stockmen's dining rooms. Most people sleep in their vehicles on the show ground. If you don't do this you will miss half the fun of showing and probably fail to make many friends. If you intend to sleep in your horse-box this must be mucked out before it gets dark otherwise the operation will be doubly difficult and the floor tends to dry off quicker in daylight. Before you turn in for the night make quite sure you know where the collecting and judging rings are, otherwise you may have to waste valuable time the next morning looking for them.

If you are sleeping in a horse-box or trailer, it is definitely advisable to have an interior light fitted, otherwise you will be groping around in the dark when you go to bed. As far as the actual bed is concerned, a camp bed can be used but if you intend to sleep on the floor in a sleeping-bag or on a mattress sufficient straw should first be placed on the floor. Most horse-boxes and trailers have corrugated floors which are incredibly uncomfortable to lie on for any length of time, but a thick covering of straw makes them quite acceptable.

At most shows the youngstock and breeding classes are scheduled to start at 9.0 a.m., so when you turn in for the night set your alarm clock for about 5.0 a.m., as an early start in the morning is essential.

When you get up in the morning and have washed and dressed, go straight to the horse lines and start by mucking out. The dirty straw should be thrown in a pile outside the loose-box door from where (at many shows) it will be collected by a gang of men with a tractor and trailer at about 6.30 a.m.

The loose-box should be bedded down well and the straw built up round the walls so that it looks tidy when the general public come around later to look at the horses. If your horse is inclined to get upset by having lots of people looking at him and stroking him, or if he is a colt and has a mare or filly in the box next to him, it is often advisable, in the case of open-fronted boxes to put a weld-mesh cage across the front of the loose-box. If he is a tall horse put a canvas or wooden screen round the top of the walls inside the box to prevent him touching noses with horses in adjoining boxes.

As soon as the horse has been mucked out, any stable stains and top dust should be removed from his coat, and his mane and tail brushed out well. Then if he is to be shown in any class other than mountain and moorland ponies, palominos or pure-bred Arabs, he must be plaited up, and this includes Arabs, palominos and mountain or moorland ponies shown in classes other than those listed exclusively for their particular breed or type, e.g. riding pony, breeding or hunter breeding classes.

To plait a horse's mane and tail you will need the following equipment:

(i) A headcollar and rope; the horse can be tied up or

if he is a youngster he would be far better held by someone;

(ii) if you are small or your horse is tall, you will need
a box to stand on;

(iii) a water or dandy brush to dampen the mane and
tail, and, of course, a bucket with some water in it;

(iv) some thick harness thread, which is obtainable
from most saddlers, and as near to the colour of the
horse's mane and tail as possible;

(v) a mane comb and scissors;

(vi) at least one blunt needle, also obtainable from
most saddlers.

Before you start plaiting your horse's mane, examine
the conformation of his neck. A horse with a slightly
ewe neck should have fewer plaits put in loosely especially
over the base of the neck, as this helps to fill up the
hollow on the top line of the neck; whereas a horse with
a short or thick neck will need the maximum number of
small tight plaits to create the appearance of a longer
neck. Horses with ewe necks therefore need to have their
manes pulled longer than the average.

The exact number of plaits you intend to put in your
horse's mane should be determined before you start, as
it is essential that every plait is made exactly the same
size right down the neck. The current fashion is to make
a large number of very small button plaits. These look
good and greatly improve the appearance of a well-
developed neck.

Thread the needle and with the thread double tie a
knot in the end. Stick the needle in your lapel and pass
the loose end of the thread over your shoulder so that
the needle does not get pulled out — there is less likeli-
hood of the thread getting caught in something if it is
hanging down your back.

First dampen the mane with water, estimate the
number of plaits you are about to put into it and calcu-
late the width of each. Make sure all the hairs are lying
over on the offside of the neck. Divide the forelock

from the first plait immediately behind the ears by making a parting with your comb; divide this first plait from the second in the same way, measuring its width with your comb.

Plait the hair right down to the very bottom and secure the end with a knot as shown in plate 24. Bring the needle back into the base of the plait, pulling the end up to form a loop. Take the needle back into the end of the loop and double the plait up again, making sure that any stitches on the surface of the plait are kept small so they don't show. Next, wrap the thread two or three times round the base of the plait and pass a stitch

24. Stages in plaiting a mane. The first is made immediately behind the ears, and each plait must be the same width with a straight parting between every one. (1) Plait the hair down to the bottom and secure the end of the plait with a knot; any loose hair should be wrapped round the bottom of the plait and secured with a stitch. (2) The long plait is doubled up and a stitch passed through the double thickness. (3) The thread is passed through the loop of the plait. (4) The plait is doubled up again and secured. (5) The plait is doubled up again, when necessary. (6) The thread is passed round the base of the plait and a stitch is passed through the thickness of the plait and secured on the underside. Always use a blunt needle when plaiting up horses.

right through the thickness at the base of the plait. Make sure that you catch every layer of the plait. Knot the thread on the underside and cut the thread short enough so that it doesn't show under the plait.

Keep the mane damp and continue to plait down to the withers. The last plait or two can be made a little slacker to allow the horse to stretch out his neck without it pulling too much. Once the mane has been plaited up completely, any loose hairs which are sticking up can be pulled out — on no account should the hairs be cut otherwise in a week or two they will have a hedgehog appearance and look ghastly.

Never plait foals' manes — they haven't got the neck development to look nice. A foal's tail should, however, be plaited.

If your horse's tail is not pulled, when you have finished plaiting the mane turn the horse round with its back to the front of the box and start plaiting its tail

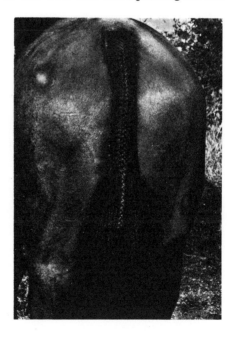

25. A plaited tail. The secret of good plaiting is to take only a few hairs at a time.

Take a few hairs from the right underside top of the tail and pass them across into your left hand.

Take a few hairs from the same area on the left side of the tail and pass these over the first into your right hand.

Introduce more hairs from the right side immediately below those already taken, using your thumb and first finger to hold the crossed-over hairs, pass the new hairs "C" over the top of "A" and 'B".

Bend "A" round over the top of "C", passing from left to right.

Introduce more hairs from the left side and add these to "A", bring "B" across the top of "A" to the left side of the tail, introduce more hairs from the right to join "B" and repeat down the tail

Fig. 23. Stages in plaiting a tail.

(see plate 25). Make quite sure that there are no knots in the tail, then dampen the top of the tail with water and commence plaiting as high up the tail as possible. Take up three strands at the top of the tail, introduce a further strand from the right side underneath the tail and then from the left side underneath the tail (see

Fig. 23). Continue down the tail until you are about an inch above the end of the dock; if the horse has a very short tail, as in the case of foals and some yearlings, a plait of considerably shorter length should be made. When you get to the required length, stop introducing hairs from the underside of the tail and simply plait the hairs in your hands down to the end to form a long pigtail. The end should be secured in the same way as the mane plaits. The pigtail can then be looped up and its end tucked into the plaited part of the tail and secured with small stitches. The secret of good tail plaiting is to take only a few hairs from the undersurface of the tail each time otherwise the plaiting will not be tight. For a more interesting effect the pigtail can be rolled up and secured with a few stitches at the end of the plaited part of the tail.

When you have plaited the tail, put on a damp bandage, which helps to flatten the hairs and greatly improves the appearance of the plaiting. When you take off the bandage, just before you go to the collecting ring, do not pull it off in the usual way as this would damage the plaiting; instead the bandage should be very carefully unwrapped. Do not leave a bandage on a foal's tail for more than an hour at a time.

In the case of horses shown unplaited, the mane and tail should be brushed out thoroughly and then combed through so they are completely free from knots. With mountain and moorland ponies shown in hand unplaited, in the breed classes, it is customary to plait the section of mane immediately behind the ears, so that it stays in place tidily and does not get mixed up with the forelock. The plait is left as a narrow pigtail and is not turned up in the usual way.

When you have finished plaiting up, refill your horse's water bucket, give him a feed and go and have a wash and brush-up yourself before breakfast. Immediately after breakfast collect your ring numbers, if you have not already done so, and where necessary take your animal to be measured. All horses entered in height-restricted classes which do not possess current height certificates must be measured. Some shows measure all youngstock entered in height-restricted classes, regardless of any documents they may possess. All brood mares are examined for hereditary unsoundness.

Next give your horse his final brush over. If he is normally out at grass all the time, a very satisfactory result can be obtained by running a damp body brush over his coat always in the direction in which the hair grows. As soon as it dries, shine the surface with a coarse stable rubber — this method does not get rid of the natural grease in the horse's coat, which he will need when out at grass, it only shines up the surface of the coat and is quite sufficient for showing brood mares and youngstock.

Any white socks or stockings should be rubbed over with a block of dog chalk which removes yellow stains and adds brightness. The hooves are then oiled.

When the horse is quite ready, the person who is to lead him in the show ring should get washed and changed, and the correct ring number must be tied round his or her waist — the only time these numbers are put round the horse's neck are for heavy horses and carriage horses, e.g. Cleveland Bays, Shires, etc. At some shows it is customary to tie the ring numbers to the bridle, in which case specially designed numbers are provided.

The show bridle and rein can then be put on the horse and the tail bandage removed. Wait in your loose-box until the first call is made over the loud speakers or by the horse foreman for the entries in your class to go to the collecting ring; make your way slowly, there will be plenty of time. The collecting ring stewards will tell you when to go into the ring for judging. If you have never shown a horse in hand before it would be better if you allowed one or two seasoned exhibitors to go in front of you and give you a lead, otherwise it is a good idea to lead the class in yourself, as the judge is bound to look at the first horse which comes into the ring even if he doesn't like what he sees. Otherwise, if it is at all possible place your horse between two obviously inferior ones so that the comparison is in your favour.

Walk out smartly on the right rein. If the horse in front of you is walking slowly go on a slightly wider circle but don't let your horse be obscured from the judge's view.

The judge will now be assessing his class and will soon start to pull the horses he likes best into the centre of the ring. Keep half an eye on the judge and be ready to come into the middle of the ring should the ring steward ask you to. When you are pulled in, stand your horse up facing the grandstand and make sure he is standing with his weight evenly distributed on all four legs. The judge may come up to each horse and examine it while it is in line, otherwise he will examine it when it is pulled out of line in front of him. Wait until the ring steward tells you to bring your horse out for the judge to see, then stand the horse in front of the judge and preferably sideways on to the crowd. Make sure he is standing with his weight evenly distributed on both front legs and with the hind leg furthest away from the judge slightly forward.

When the judge has finished looking at him he will ask you to walk to the far end of the ring, turn your horse and trot back past him to the other end of the ring. As you go up the ring make sure your horse is really walking out to the best of his ability. As you turn, walk round your horse, never ask the horse to walk round you otherwise he will tend to become unbalanced, also you may lose control. The horse must also be balanced when he is asked to trot on; therefore when you turn to trot back up the ring, walk your horse for at least three strides before giving him the command to trot on. Be ready to give him a tap with your stick if he fails to get the message.

Providing you have a horse which moves really well, it

is better to trot out fast and risk your horse breaking into a canter than to jog up slowly. This way the judge can see before your horse breaks that he is an outstanding mover. Once your horse is trotting on well, he may, as some horses do, 'change gear' as he lengthens his stride. Drop your hand so that the horse is going on a slack rein and he will have to carry himself and not lean on the bit with his weight on his forehand and his hind legs trailing behind. Some Arab and pony judges require you to trot once round the ring but hunter judges require only that you go the length of the ring. Just before pulling up give the horse the command to slow down as, if obeyed, this will obviate the necessity of hauling on his mouth.

After your show, return to your place in the line and be ready to stand your horse up again should the judge come back down the line examining the horses. At this stage most judges ask the whole class to walk round the ring again in a somewhat reduced circle, then they start to call the exhibitors into their final order of judging and the rosettes are handed out.

It is only polite to thank the judge as he hands you the rosette and card, even if you are disappointed or have only received a commendation.

When you have finished being judged and before returning to your loose-box make quite sure you are not wanted back in the ring to compete for a championship or special prize -- all first and second prize-winners from every class, except for foals, are usually required to compete for the breed championship. If there is more than one foal class, foals are often required to compete for their own championship.

The foregoing description of the method of showing horses in hand applies directly to youngstock and not specifically to mares and foals. In their case the ring procedure is identical except that the mare and foal are always shown in the ring together and must be led round the ring by two people, one leading the mare and the other the foal. The mare is usually led in front of the foal and judged first. At some shows the foals are judged immediately after the mares, while at others all the mare classes are judged first and the foals from all the classes are divided into their sexes and each sex, i.e. colts or fillies, is judged separately.

When you are called into the middle of the ring, whichever animal is being judged must be made to stand in front of the one which isn't being judged. When they are trotted out they should be trotted side by side and not behind one another, otherwise the judge will be unable to see clearly the action of the one he is trying to judge.

All prize-winners are bound to parade at every show where a grand parade of prize-winners is staged — failure to parade usually means withdrawal of any prize money won. A steward is posted on the collecting ring or main ring gate to take down the numbers of all exhibitors who parade. This grand parade is usually held in the late afternoon of the day of judging, or at some shows on the next day. This means that if you have won a monetary prize you must stay at the show for another night, in which case there is usually no extra charge for stabling.

It is customary not to plait up for a parade and indeed most people remove all their plaits before the parade so that they can load up and go home immediately afterwards. Non-prize-winners are usually at liberty to go home straight after judging, although some shows keep their gates shut until about 4 p.m. so that the general public can see all the horses.

If you have won a major prize, after the judging a steward may come round to enquire about your horse's past show record. This information is for the commentator during the grand parade and will be read out as your horse enters the ring.

When you are getting ready for the parade, as far as possible fix your horse's rosettes to the nearside of the bridle, so that the crowd can see them as you walk round. Any ribbons which could flap into the horse's eye should be anchored underneath the bridle, so that they cannot frighten him. All rosettes from every class won at that particular show should be worn for the parade, otherwise during the judging only rosettes won in the particular breed or type class should be worn, regardless of whether you are still showing under the same judge.

About half an hour before the time to parade you will be called down to a collecting ring, where you will be sorted out into class order; the champion and reserve champion from whatever class they came, usually lead the parade of each breed or type followed by the prize-winners in order starting with the first prize-winner of each class.

The prize-winners usually walk once round the main ring and then line up across the middle of the ring. The cups and trophies are then awarded and the winners go forward to receive them. After the presentation, the prize-winners file once more round the ring and then go out.

Small shows and gymkhanas differ in some respects from the large shows. They seldom have veterinary surgeons on the gate to check the health of the horses coming onto the show ground. However, they do have an official on the gate to collect entrance money from the general public and direct exhibitors.

Your horse-box should be taken to the parking space allotted to these vehicles. This type of show seldom, if ever, provides stabling for its exhibitors; therefore the horses must remain inside the horse-boxes when not actually being shown or being prepared for the ring.

Quiet horses can be tied to the side of the horse-box with a hay net to keep them occupied, but this is not to be recommended for youngstock.

On arrival at the show, it is advisable to go to the secretary's tent and make your entry, if you have not already done so, and collect your ring numbers.

Most larger shows state that foals must be at least one month old before they can be shown. Smaller shows tend to leave it to the good sense of their exhibitors and foals which are barely a week old sometimes appear in the ring. This practice is not to be recommended for two good reasons:

(i)   the foal is too young to be asked to walk round and round a show ring, while first the mares and then the foals are being judged; it will get tired and this lowers resistance to disease;

(ii) by taking a very young foal to a horse show you are exposing it to diseases to which it may have no immunity and could therefore become ill.

For one-day shows it is absolutely essential to arrive at least an hour before you are due in the ring. If your horse will stand quietly while you plait it up at the show, it is a good idea to leave the tail until you arrive on the show ground, as tails can easily be rubbed out while the horse is travelling. The mane is probably better put in the night before the show or early in the morning before you leave home. There are two disadvantages to plaiting up manes the night before the show: (i) horses tend to rub their manes and necks when they are left plaited up overnight, so even if the plaits don't actually come undone, they will look rather wispy by the morning; (ii) small pieces of straw tend to get entangled in the plaits when the horse lies down during the night, and these have to be picked out carefully in the morning. However, if plaiting up is left until arrival at the show ground, it may mean a very rushed job, as there can be traffic hold-ups on your way to the show.

At small shows it is unwise to rely on an official to tell you when to come into the ring for judging, so keep a very close watch on the progress of the ring events otherwise you might miss your class. Some small shows do not have a collecting ring and the entries file straight into the judging ring. Many small shows have separate showing rings for in-hand and ridden classes. The ridden classes and show jumping always take place in the main ring, whereas the in-hand classes often take place in a subsidiary ring.

The turn-out and method of judging and showing at small shows is identical to that already described for larger shows. Small shows don't usually organise a grand parade later in the day, so once the judging is over the exhibitors are usually free to go home.

As far as many of the larger shows are concerned, where loose-box charges are not included in the entry fees, horses may be brought to the show ground on the day of the show -- the latest time of arrival is usually given in the schedule.

13 The Sale of Brood Mares and Youngstock

There are two ways of disposing of animals you have for sale. One is to sell them privately, the other is to take them to a horse sale.

Selling privately is to be recommended wherever possible if you wish to have some control over who buys the horse. To sell privately an advertisement in the local press, and possibly also in a horse magazine or paper, will usually have the desired result. Alternatively, if the horse or pony in question is a top-class show animal, there are often several people only too willing to buy following a success in the show ring. It is common knowledge that you should never turn down a good offer for a horse, as if you do something will surely happen to the animal, and you should never buy or sell a horse on a Sunday, as that too brings bad luck. Some breed society shows advertise horses which are for sale by organising parades for them at their annual show or putting the words 'For sale' under the horse's name in the catalogue.

Generally speaking, preparation of youngstock for the sale ring is identical to that necessary for the show ring. Most youngstock look better if they are shown or offered for sale slightly on the fat side. A good layer of fat tends to cover up conformation faults, as it rounds off the contours and makes the majority of horses look a better shape, except perhaps for cobby ponies, which if shown too fat tend to look stuffy.

Thoroughbred youngstock, bred for flat racing, are usually broken at two years and therefore only appear in the sale ring as unbroken stock up to this age. Horses bred specifically for National Hunt racing are not broken until they are at least three years old and sometimes much later. Unbroken National Hunt youngstock normally sell best as three-year-olds.

Thoroughbred foals reared for sale must be well fed but not over-fed. Fatness at this age can lead to problems with the growth plates in certain individuals. Thoroughbred yearlings, by tradition, are brought to the sales in hard, fat condition.

It is advisable not to sell other breeds at auction sales until they are five years old and quiet to ride or drive.

Most horse sales are advertised in *Horse and Hound* throughout the year and in almost all cases a catalogue of lots to be sold is printed by the auctioneers concerned. Some of the smaller sales accept entries on the day of the sale but also produce catalogues. It is in the owner's interest to enter a horse early enough for the description to appear in the catalogue and so be read by everyone interested in the sale. Many people looking for a particular type of horse or pony, write for sales catalogues and go through them before making up their minds which sale to attend.

As far as unbroken youngstock are concerned, these are usually sold without any warranty for soundness.

Most youngstock (with the possible exception of National Hunt stores, which should be sold mouthed only) sell better as three-year-olds if they are mouthed and backed but unschooled. Children's ponies sell better at five years old or over if they are schooled and can be ridden by a child. Ponies under three years old should never be broken and ridden as even if they look adult they will lack maturity and could go lame with work.

Horses which are broken and known to be sound should be sold believed sound in wind, heart, eyes and action and open to veterinary examination.

For pure-bred animals a three-generation pedigree should be given, supported by family details of those three generations on the dam's side only. The pedigree research is carried out by agencies or the auctioneers themselves. This is of particular importance in the case of Thoroughbred racehorses which are often bought as much for their pedigree as for their conformation, although the latter is probably more important.

The seller (vendor) is ultimately responsible for the accuracy of the description and pedigree of their horse, which is now of major importance since the Misrepresentation of Goods for Sale Act of 1968.

Any horse entered for a sale which is primarily for Thoroughbred horses and which is not entered in the General Stud Book, must be so described.

All animals up to four years old must be described as colts, fillies or geldings according to their sex, and after this age as stallions, mares or geldings. The only exception to this rule is where a filly, prior to four years old, is covered by a stallion, in which case she must be described as a mare, even if she is only a two-year-old.

In the case of a brood mare the covering sire and exact date of her last service must be given, and where the mare is sold as believed to be in foal, this statement must be supported by a recent pregnancy certificate which should be given to the auctioneers before the sale.

Always read the conditions of sale carefully before entering an animal. Generally it must be declared if a horse is a rig, weaver, windsucker, crib-biter or is unsound in its wind: auctioneers' regulations do vary, however, from one firm to the next.

If the buyer of a horse which has been sold with any kind of warranty or description contends that the horse in question does not correspond to that warranty or description, he can return the horse to the auctioneers, usually within forty-eight hours of the sale, but in some cases, as long as seven days after the last day of the sale. The auctioneers or someone appointed by them will settle the dispute.

The horse ceases to be the property of the vendor from the moment the hammer drops and from that time becomes the sole responsibility of the buyer.

Most auctioneers charge an entry fee and take the guineas from the sale price, all bidding at horse sales being conducted in guineas and not in pounds.

A reserve price is the sum at which you are prepared to sell your animal. If you wish to have a 'reserve' on your horse you must declare this to the auctioneers before the sale, otherwise your lot will be sold 'without reserve' to the highest bidder.

Once a horse has been entered for a sale, normally it must not be sold privately within forty-two days prior to the sale or within seven days after the sale, unless a fee is paid to the auctioneers.

When preparing riding ponies for the sale ring the main aim is to make them look as much like Thoroughbreds as possible. There is a real art in producing horses for the show ring and sale yard but it is often possible to improve the appearance of most horses with careful trimming.

A Thoroughbred has very little, if any, hair around his heels, hooves, head or ears — the more highly bred he is, the less hair he has. The hair on his body is short and fine; his mane and tail are usually thin, straight and silky.

The common bred horse on the other hand has a bushy mane and tail which is coarse, thick, and should therefore be pulled and thinned. It will probably have a beard and whiskers. Long, coarse hairs round the eyes and muzzle can be cut off, as close to the skin as possible, with a sharp pair of blunt-ended scissors. Any fluff in the ears can be trimmed out in the same manner, but be careful not to let hair fall down into the ears, as this might set up an irritation which could cause head shaking. If there is only a small beard under the chin, this can be removed by singeing; but a thick beard and any hair in the heels should be plucked out or removed with clippers — it must never be cut with scissors, as this leaves ridges.

To trim surplus hair with clippers, insert a coarse blade into the clipper head and screw down the head, but not tightly. Apply plenty of '3 in 1' oil to the head and along the edge of the blades. Start the clippers working and allow the horse time to get used to the noise before you start trimming. Then run the clippers down the coat in the direction of the hair — as opposed to clipping a horse when you always run against the coat. This takes the long hair off without making the horse look obviously clipped. Any hair round the top of the hooves can be removed either with the clippers or with a sharp pair of blunt-ended scissors.

Trimming should be done at least a week to ten days before a horse is going to appear in public, to allow the hair time to grow a little and look more natural. White markings are usually stained slightly yellow on the surface and therefore show pure white in patches for several days after the hair has been trimmed.

A long, coarse winter coat can be made to look better if the horse is trace clipped, but where it is known that the animal will be kept inside and rugged up, a hunter clip may be given leaving only the saddle patch and the legs. A horse or pony which grows a coarse summer coat can be made to look better if it is kept rugged up all the summer and/or if the coat is clipped in the early spring just as the new summer coat is coming through; this is an exception to the general rule and if practised on all horses would ruin the summer coat of more finely bred individuals.

Breeds of horses and ponies which are normally shown unplaited and virtually untrimmed, i.e. in their natural beauty, and this includes palominos, should be taken to a sale in the same way. Their tails should only be pulled at the bottom. Other horses should be plaited up and their tails pulled or plaited and cut level to just below the hock joint for horses, or above the fetlock for ponies. The tail should be cut so that it slopes slightly upwards towards the horse's hocks and not straight across. This way it will appear level when the tail is carried naturally. Thoroughbred horses entered for sales exclusively for this breed, except for those actually in or just out of training, are always sold unplaited, but they should have their manes neatly pulled and laid. At ordinary horse sales and fairs they should be sold plaited up.

The tack used for sales, and indeed the standard of turn-out for both the horse and the attendant or rider, should be the same as for the show ring.

One of the most important things to remember about horse sales is to arrive early, so that prospective buyers have plenty of time to examine your horse before it goes into the sale ring. Most of the larger sales have stabling available for horses and it is a wise plan to make use of it. Some buyers come round the sale yard the day before the sale to view the lots.

At the larger sales, auctioneers come round to inspect the lots they will sell, meet the vendors and discuss reserves. Additional information which may help to sell the horse should be volunteered. At smaller sales and horse fairs it is customary to stand up by the auctioneer's side while he is selling your animal and to give him any information concerning the horse together with the reserve price. At these sales you can then agree to accept a lower price if the bidding does not reach the original reserve, before the animal leaves the ring.

At all sales each horse is issued with a lot number. These are usually printed on small, oval paper discs which are stuck onto the horse's rump, one either side. Sophisticated plastic discs can be hung round the neck or numbers tied to the bridle each side — in which case make sure they are on the right way up as nothing looks worse than a number tied on upside down.

At the smaller sales, the horses usually come into the ring on the auctioneer's right and stand in front of him while their description is read out: this includes their age; sex; height; whether they are broken and quiet to ride and/or drive. At some sales unbroken stock are sold in a separate ring. Details of any prizes or competitions won by the horse are usually given.

The animal is then turned round and trotted out smartly to the end of the ring, turned and trotted back. This is usually repeated at the request of the auctioneer if there is a pause in the bidding. The usual dealers' term for trotting a horse out in hand is 'give us a yard'.

At the larger sales, the animals are walked round in a preliminary ring before they go into the pre-sale ring. They can be pulled out and inspected by anyone who wishes to do so. At all times horses must therefore be walked out smartly at an energetic pace. A totally unfit horse soon tires, therefore some preliminary sales preparation is essential. Thoroughbred yearlings are usually walked for an hour a day for about a month before the sale. In the sale ring itself the horse must be walked round very smartly on the right rein – it will not be required to trot.

All lots must be paid for before they leave the sale yard. Payment to the vendor for lots sold is usually made by the auctioneers automatically a few days after the sale but some firms require a written request for payment from the seller or his representative, before any money changes hands. It is therefore just as well to read the conditions of sale in the catalogue.

14 Disease Conditions of Young Foals

Unlike the human baby the foal does not receive any resistance to disease before birth and has to acquire all its immunity from the colostrum in the first twenty-four hours after birth. The colostrum or foremilk contains high levels of antibody and the ability to absorb these across the small intestine ceases before thirty-six hours of age. Antibodies are substances found in the blood which destroy germs that would otherwise cause disease. After thirty-six hours the foal can only obtain immunity artificially by means of injections or by natural development over a period of time.

When dealing with the problem of disease, prevention is always better than cure and is far cheaper in the long run. Wherever possible I will point out the preventive measures available to the owner for all the more common conditions and diseases met in young foals, as well as the symptoms and when to call in your veterinary surgeon.

A summary of disease conditions in foals is given in Fig. 28 at the end of the chapter (see page 161).

BARKERS, WANDERERS AND DUMMY OR CONVULSIVE FOALS

These terms are used to describe obscure conditions sometimes seen in newborn foals. They are attributed to brain damage caused by lack of oxygen at the time of birth, or to infection by the germ *Actinobacillus equuli* or other organisms such as those which cause septicaemia.

Barker foals will often appear perfectly normal at birth but some time afterwards may start to make a noise like a dog barking, and will show marked signs of respiratory distress and convulsions. If the foal is already lying down, it may jerk its head up and down and move its legs without being able to get up. If it is already on its feet it will probably show symptoms of blindness: wandering around aimlessly, bumping into the walls and continually calling for its mother without apparently being able to recognise her.

The foal should be placed on a blanket to prevent any bedding getting into its eyes, which at the very least would irritate them, and at the worst cause blindness. The blanket should be placed under a heat lamp for warmth and an old sweater can be put on the foal. Veterinary aid must be called in immediately before there is permanent brain damage.

Lack of oxygen at the time of birth, associated with premature breaking of the cord, or fractured ribs sustained at the time of birth, can cause these conditions.

Wanderers and dummy foals are variations of the same condition. Wanderers, as the name implies, describes a foal which walks aimlessly round the loose-box, showing definite signs of irritability and restlessness. It will go off suck, chew incessantly or grind its teeth. Dummy foals on the other hand lose consciousness and become comatose.

These foals must be fed by stomach tube if they are to have any chance of recovering. This condition is more common in Thoroughbreds.

DEVELOPMENTAL ABNORMALITIES (plates 26–28)
Foals, like babies, can be born with various weaknesses and deformities including 'parrot mouth', cleft palate, hernia and deformed limbs. Some of the more common limb deformities are:

1 *Over-long weak pasterns:* In these cases the foal walks on its heels with his fetlocks touching or almost touching the ground. The condition usually rights itself as the foal becomes stronger. In most of the milder cases of limb weakness in young foals, exercise out in the paddock will help to strengthen their legs.

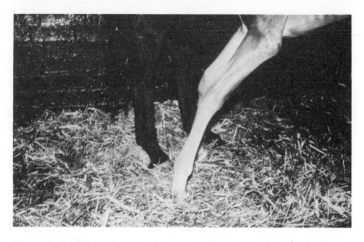

26. A foal with contracted tendons, showing maximum extension of the front legs. This foal came right in two months, after splints had been applied, and he eventually raced.

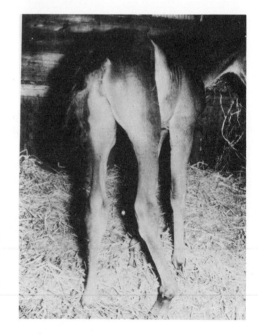

27. Another foal showing a contracted tendon behind. With the help of support bandages this condition was corrected within six weeks of birth.

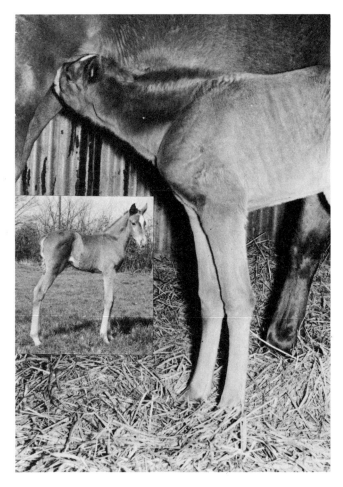

28. A foal born with very weak front legs causing it to stand right back at the knee. Inset: the same foal two weeks later.

Since the heel and/or fetlock are touching the ground they will become very sore if protective bandages are not applied. Care should be taken that the joint does not become infected, otherwise an infective arthritis (joint ill) might result. Once the fetlock has come up off the ground the bandages may be dispensed with, but any sore places should be treated with an antibiotic spray, some of which also have the effect of hardening the skin.

2 *Contracted tendons:* The foal may only be able to walk on the tips of his toes, or in very bad cases be entirely unable to get up. If the contraction is very bad the foal may have to be put down, but in some cases an operation to cut the tendons and straighten the leg may be successful. In other cases splinting the leg may work well.

3 *Twisted or bent legs:* When foals are born with twisted legs which do not start to straighten within a week of birth, or where their legs start to twist soon after birth, your veterinary surgeon should be consulted, as the longer these weaknesses are left the harder they are to correct. Correction must take place while the foal is still growing, and may take the form of plaster casts. In some cases a leg which is growing unevenly may need to be stapled to straighten it but this is a major operation so the value of the foal must be taken into consideration. (See Fig. 24.)

Feet which turn in or out can usually be corrected quite satisfactorily by the blacksmith. Any minor deviation is rectified by careful trimming but a major fault will probably need special shoeing from the moment the foal's hoof can take a shoe (usually by eight weeks old, although shoes which are stuck on may be applied

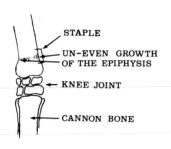

Fig. 24. Uneven growth of the epiphysis.

STAPLE

UN-EVEN GROWTH OF THE EPIPHYSIS

KNEE JOINT

CANNON BONE

earlier). The outside of the shoe is built up to turn the toe out, and the inside built up to bring the toe in. More slip can be achieved if the front of the shoe, at least, is smooth. Corrections to any joint must be made gradually, otherwise an enormous strain will be placed on the affected limbs. When the ground is soft, these animals should be walked on a hard, level surface for a minimum period of half an hour each day.

Developmental abnormalities may be inherited or they may be due to environmental conditions.

DIARRHOEA OR SCOURING
This can be a symptom of several diseases, errors in feeding or worm infestations.

Most foals will suffer from diarrhoea at some period before they are weaned, often when the mare comes in season after foaling. This is thought to be due to changes in the composition of the mare's milk at this time. But the ingestion of poor-quality high-fibre feed, such as inferior hay, before the foal's digestive tract is capable of dealing with these substances, may play a part. The foaling heat scour usually stops quite naturally soon after the mare has ceased to be in season. The foal does not go off suck or show any other symptoms of illness but this 'natural' scour can weaken the foal and make it more susceptible to infectious scours at this time. Dosing with kaolin mixtures such as 'stat', two or three times a day usually controls non-infective scours.

When foals scour, the tail and adjacent areas should be washed at least once a day and some Vaseline should be smeared round their hindquarters, otherwise the area will soon become raw and bald. This is especially important if you intend to sell or show your foal before weaning.

All infectious diarrhoeas should be isolated and have immediate veterinary attention, as young foals have very little reserves and are soon pulled down in condition.

When a foal is scouring there is usually a character-istic odour which is immediately recognisable on enter-ing the loose-box. The foal's temperature is usually raised. The thermometer must be held against the side of the rectum otherwise, in cases where the rectum is relaxed, a false reading might be obtained. The foal may go off suck, or may continue sucking. Some foals drink large quantities of water and their stomachs become dis-tended. Foals which stop sucking soon become dehydr-ated, and in these cases a balanced electrolyte mixture such as Ionalyte should be given by tube. This can also be added to the water of foals that are drinking readily. On no account should water be restricted. Signs of de-hydration usually include sunken eyeballs, staring coat and a weak pulse.

Draughts and sudden changes in temperature can be predisposing causes of scouring. When it occurs the most important thing to remember is that any fluid loss must be replaced and the electrolyte balance maintained. If the fluids are not replaced in adequate amounts the foal will die as a direct result of fluid loss. The foal's tongue will often become coated and there is a marked loss in condition and wasting of the muscles.

Injection of some of the mare's blood into the foal will help in some cases due to the antibodies this blood will contain. Antibiotic treatment is usually necessary for recovery.

ENTROPION
This is a condition in young foals which produces a chronic watering of one or both eyes. This becomes ap-parent as early as one or two days after birth. On careful examination it can be seen that the bottom eyelid has turned inward and the lashes are irritating the eye, causing inflammation of the cornea.

This is a job for your veterinary surgeon and he may wish to insert some stitches into the bottom lid, which will have the effect of making the lid turn outwards in

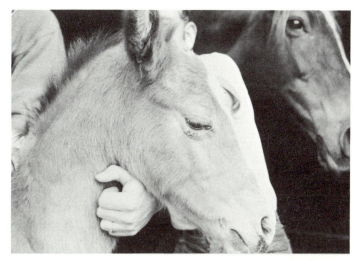

29. A foal showing the treatment for entropion.

the normal manner. The stitches are usually left in place for about two weeks (see plate 29).

EQUINE INFLUENZA

Affected horses will cough, run a temperature and sometimes have a slight nasal discharge. Adult horses usually recover in about a week, but young foals often become seriously ill and some die, as they have very little resistance to influenza at this early age unless the mare has been vaccinated recently before foaling.

The incidence of influenza among foals can be greatly reduced by ensuring that all brood mares are vaccinated annually and that the booster vaccination is carried out in the last month of pregnancy. This will produce the maximum amount of antibody in the mare's colostrum. No ill effects as a result of vaccination, either to the mare or the foal she is carrying, have ever been recorded. Foals should be vaccinated at three months and again at four months, with a booster dose twelve to eighteen months later.

As far as Thoroughbred yearlings are concerned, it is recommended that vaccination should take place in July or August.

HAEMOLYTIC DISEASE

This is a blood disease of the newborn foal in which the chief symptoms are anaemia and jaundice. The disease is caused by a blood group incompatibility between the foal and its dam, and is somewhat analogous to Rhesus babies in human medicine.

The condition arises when a foal, while still in the womb, has inherited a blood group from its sire which is incompatible with the dam's blood. The mare then becomes sensitised to this 'foreign' blood group and produces antibodies against it. She concentrates these antibodies in her colostrum so when the foal first sucks they pass into its bloodstream. The antibodies then combine with and destroy the foal's red blood cells, which results in severe anaemia and jaundice.

If you own a mare which has previously had a foal with haemolytic disease or if you want to make sure that your mare does not have a foal with haemolytic disease, then about three to four weeks before she is due to foal ask your veterinary surgeon to take a sample of her blood. This sample will be laboratory tested and compared with samples taken during the following two weeks to see if there is any increase in antibody level. Any significant increase would denote that the foal will be at risk should it receive colostrum from its own dam.

If it has been determined that the mare has an increased antibody level and that the foal when it arrives will therefore be at risk, the main preventive measures to be taken are:

1 Obtain a foal-size muzzle — this must be put on the foal at the moment it gets to its feet, to prevent it from drinking any of its mother's milk and must be left on

for thirty-six hours. If all access to the dam's colostrum is withdrawn the disease cannot occur.

2 As the foal must have some colostrum an alternative supply must be provided. This can be obtained at any time from another unsensitised foaling mare and kept in a deep-freeze until needed. It should be warmed to blood heat before feeding. For a thoroughbred foal allow six fl. oz every three hours for twelve hours, less pro rata for smaller foals. The mare must, of course, be milked out at least six times a day to prevent her from drying up before the foal is allowed to suck. This milk must be thrown away.

The clinical signs of haemolytic disease are that the foal will be seen to yawn, appears sleepy and short of breath due to lack of oxygen in the blood, which in turn is due to destruction of the red blood corpuscles. The mucous membranes are yellow. This condition usually becomes apparent within the first thirty-six hours after birth.

HERNIA

A hernia or rupture is caused by incomplete development

Fig. 25. Hernia.

of a muscle wall (usually abdominal) and the protrusion of an organ or its fatty connective tissue through the hole which has been formed, causing a swelling to the outside beneath the skin. Umbilical hernias are the type most commonly met in foals.

They usually appear at about six weeks old and are seen as a swelling in the naval region, due to part of the bowel protruding through the umbilical ring which has failed to close up in the usual way.

This type of hernia should not be confused with thickening which sometimes takes place in the region of the navel which is hard and unyielding. In the case of a hernia a distinct ring can be felt through which the protruding mass can be pushed back into the abdomen and through which it protrudes again once the pressure is released.

An umbilical hernia is sometimes self curative and tends to disappear quite naturally as the animal gets older, but in some more severe cases surgical treatment to close the hernia ring is necessary. This should be done before weaning at about four months old.

Some people advocate applying blisters to the area with some success but this tends to thicken the skin making surgery if necessary more difficult.

JOINT ILL OR NAVEL ILL
This condition usually becomes apparent some time between five days and four months of age, but most commonly at about three weeks. It is characterised by swellings in the leg joints or around the navel. The foal usually goes lame in the affected limb or limbs.

The germs which cause this disease are thought to gain entry at the time of or soon after birth. Maximum cleanliness should therefore be observed at foaling time.

The foal may go off suck and appear dull and listless, the temperature rising to $103^\circ - 105^\circ$ F. It will tend to spend most of its time lying stretched out flat on its side. One or more leg joints may appear swollen, tense and hot. If left untreated these will eventually burst, discharging bloodstained material, and the foal will gradully get weaker and may die. Sometimes abscesses are also found on the internal organs, in which case death usually occurs.

Prevention mainly consists of maintaining scrupulously clean surroundings at foaling time and a liberal application of antibiotic powder to the cord the moment it breaks. This should be repeated an hour or two later. Make sure the foal receives some colostrum within the first twenty-four hours of birth, particularly in cases where the mare has been running her milk before foaling. The mare's colostrum and foal's serum can be checked for antibody level. As a final precaution the foal should be given antibiotic cover for three consecutive days after birth.

Should you suspect that your foal has joint ill, call your veterinary surgeon immediately. If left untreated, there may be some permanent damage to the foal's joints.

MECONIUM COLIC

Meconium is the brown or black fairly hard excretory material which collects in the bowels before birth. It should be passed within a few hours, helped by the purgative action of the colostrum. The meconium is thought to consist of amniotic fluid cells, bile and debris from the intestines which have collected before birth.

It is very important that the meconium should be passed within a few hours of birth in order that normal digestion can take place. If the meconium is retained, colic will occur. The foal will probably carry its tail higher than normal, and may be seen standing in a crouched position with its back arched, straining. In a more advanced stage it will show definite signs of colic: kicking at its stomach, rolling and even lying on its back with its legs tucked up against its stomach in an effort to relieve the pain.

Retained meconium and constipation in the foal may be due to a costive diet and lack of exercise in the mare.

A dose of liquid paraffin by mouth or an enema of warm soapy water does sometimes help to relieve this condition, but it is probably better to call in your veterinary surgeon as soon as you notice the first symptoms, rather than wait until the foal gets worse. Routine dosing of all newborn foals with liquid paraffin by mouth can help to prevent meconium colic.

This condition appears to be more common in colt foals than fillies, especially overdue colt foals.

MENINGITIS

This is the term used to describe inflammation of the membranes covering the surface of the brain and/or spinal cord.

The affected foal usually goes off suck and appears generally ill, with temperature, respiration and pulse higher than normal. It may become dazed and walk into objects in its path, and develop convulsions.

Veterinary help should be called in immediately as this is a very serious disease and only very careful nursing can effect a cure.

PERVIOUS URACHUS

In the case of this condition a continuous trickle of urine will be seen coming from the navel within the first four days. It will probably take the hair off the lower regions of the hind legs, unless these parts are smeared with Vaseline to protect them.

Before birth the foal lies inside a sac known as the amnion and is immersed in amniotic fluid. This is surrounded by the chorio-allantoic membrane, which contains allantoic fluid, i.e. the foetal urine which has collected before birth. The foal's urinary bladder opens directly through the urachus into the chorio-allantoic sac, which gradually fills with urine, thus preventing over-distension of the bladder. At birth the umbilicus normally closes.

In the case of pervious urachus there is still direct access for the urine from the bladder through the navel cord to the outside. The condition will sometimes cure itself, but if it does not seem to improve within two weeks your veterinary surgeon should be called as treatment may be necessary.

PNEUMONIA

This is inflammation of the lungs with a resulting breakdown of the lung tissue which greatly impairs respiration. Respiration is faster and deeper than normal. A variety of germs, bacteria and viruses are able to cause pneumonia in susceptible foals.

Predisposing causes of pneumonia may be exposure to cold, wet weather, or stuffy and badly ventilated stables. It can also be produced by careless drenching – allowing some of the fluid to go down into the lungs, which is always a risk.

Possible symptoms of pneumonia are a high temperature and fast, laboured respiration. The foal will go off suck and there may be fits of shivering and nasal discharge. Careful nursing and management is all-important as a relapse can take place at any time if the standard of care is relaxed.

Veterinary aid must be called in right at the beginning if a good chance of recovery is to be expected.

Summer pneumonia: This is a highly infectious disease of young foals from about one to four months old, caused by the germ *Corynebacterium equi.* As its name implies this condition is most often seen during the summer months. It is more commonly found in Scandinavia and Australia but outbreaks do occur from time to time in the British Isles.

The germ causes areas of pus (abscesses) to form in the lungs. By the time symptoms of pneumonia develop the disease is usually fairly well advanced; hence the high death rate.

Antibiotic treatment by your veterinary surgeon and

good nursing give the only possible chance of recovery. Foals with pneumonia of any type must be isolated immediately as summer pneumonia can spread very rapidly to other foals.

RUPTURED BLADDER
This is a rare condition, probably due to some inherited weakness in the bladder wall.

The symptoms are similar to those of meconium colic (see page 155) and usually become apparent around the third day. The foal's stomach becomes distended with fluid and it may only be able to pass very small quantities of urine.

Your veterinary surgeon should be called in immediately this condition is suspected, as an operation will be necessary to repair the tear in the bladder wall.

SEPTICAEMIA
In this disease, bacteria have somehow managed to gain entry into the foal's bloodstream, where they circulate round the body and invade any or all vital organs. A predisposing cause of this disease can be lack of colostrum and thus the antibodies against disease which it contains. In most cases the affected animal dies; the course of the disease is often very short. It usually occurs within the first four days of life.

Sometimes the only symptoms noticed are a high temperature and general signs of illness followed soon afterwards by death. The earlier a veterinary surgeon can be called in, the more chance he will have of saving the animal's life.

SLEEPY FOAL DISEASE
This disease is caused by the germ *Actinobacillus equuli*, and normally becomes apparent two to four days from birth.

The affected foal usually appears dull and may go off suck; the temperature may be sub-normal. After several hours the foal appears sleepy — hence the name — may

show signs of colic and diarrhoea or develop convulsions.
This organism usually attacks the kidneys and brain.

The sooner veterinary aid is called, the greater is the foal's chance of survival.

STRANGLES

This is a disease more commonly seen in young horses up to five years old. It is caused by the germ *Streptococcus equi*, is very highly contagious and will cause the death of young foals. Incubation is usually from two to fourteen days. Infected horses show a rise in temperature, appear dull and go off their feed. They often have a cough, sore throat, profuse discharge from the nostrils and swellings in the angle of the jaw. This causes the horse to hold its head and neck out stiffly. After a few days an abscess will form in the swelling, mature, and finally burst, discharging bloodstained pus. This pus and the nasal discharge is highly infectious, so very great care should be taken that anyone handling an infected horse does not come in contact with other unaffected horses at this time. As soon as the abscess has burst the temperature usually becomes normal and the swelling subsides. Recovery is usually complete in a few weeks.

Treatment consists of immediate isolation of the infected animal and calling in your veterinary surgeon, who may wish to treat the animal with antibiotics.

TETANUS (LOCKJAW)

This disease affects horses of all ages but its incidence among young foals can be reduced to a very large extent by giving a mare which has already had primary tetanus toxoid injections a booster injection about three to four weeks before foaling. She will then produce antibodies which will be transferred to the foal in the colostrum, giving it immunity to tetanus for the first few weeks of its life. Immunity can be maintained up to about three months of age by giving an injection of anti-serum at six weeks old.

This disease is caused by a germ which can survive for

long periods in a dormant state outside the human or animal body. It cannot live in the presence of atmospheric oxygen. The germ gains entry through deep wounds, such as in the feet, deep puncture wounds or through the navel at the time of birth, i.e. any site not immediately in contact with the atmosphere.

The germs do not leave the site of entry but produce a toxin which they release into the bloodstream. This acts on the central nervous system producing symptoms of paralysis.

The horse shows unusual signs of nervousness. He will stand with his limbs stiff, tail raised stiffly, head and neck stretched out and ears pricked. If the head is raised quickly the third eyelid (haw) will flick across the eye very noticeably. If tetanus is suspected your veterinary surgeon must be called immediately.

Tetanus serum is faster acting than tetanus toxoid but provides immunity for about four weeks only. It has fewer side effects than the toxoid and is consequently invaluable for administration to young foals born to mares which have not themselves been injected with tetanus toxoid. It is also used after accidents where animals have cut themselves and are not immune to tetanus.

The toxoid is slower acting and is used to give lasting immunity. It is given at three months old with a second injection one month later; this is followed by a booster injection a year after the last injection to give life-long immunity. Where an animal is at risk, for instance in the hunting field, an annual booster injection is often a wise precaution.

Young horses should always be injected against tetanus as they are far more prone to accidents and injuries than older horses.

TYMPANY OF THE GUTTURAL POUCH
This is a very rare condition sometimes met in foals. The affected animal appears as in Fig. 26 with a large swelling

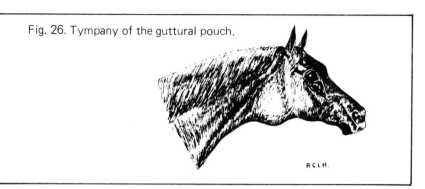

Fig. 26. Tympany of the guttural pouch.

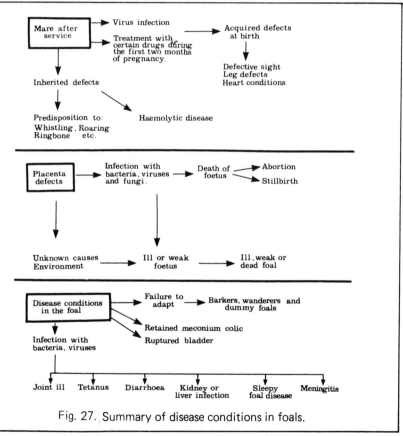

Mare after service
→ Virus infection
→ Treatment with certain drugs during the first two months of pregnancy.
→ Acquired defects at birth

Defective sight
Leg defects
Heart conditions

Inherited defects

Predisposition to: Whistling, Roaring Ringbone etc.

Haemolytic disease

Placenta defects
→ Infection with bacteria, viruses and fungi.
→ Death of foetus
→ Abortion
→ Stillbirth

Unknown causes Environment
→ Ill or weak foetus
→ Ill, weak or dead foal

Disease conditions in the foal
→ Failure to adapt → Barkers, wanderers and dummy foals
→ Retained meconium colic
→ Ruptured bladder

Infection with bacteria, viruses

Joint ill   Tetanus   Diarrhoea   Kidney or liver infection   Sleepy foal disease   Meningitis

Fig. 27. Summary of disease conditions in foals.

in the region of the guttural pouch. When pressure is applied to the swelling, air is driven out but the swelling returns within a few hours.

The affected animal makes a roaring or snoring noise on breathing and will probably make more noise when eating. In hot weather the animal will be seen to have great difficulty in breathing. There may or may not be a nasal discharge. This condition may be inherited or it may be due to an infection. Immediately you see the first symptoms, you should call in your veterinary surgeon and he may advise an operation. On no account should you attempt to lance the swelling yourself to let any fluid or air out, as this could produce a serious infection. Nowadays this condition is usually curable.

15 Disease Conditions of Foaling and Lactating Mares

BRUISING AND LACERATION OF THE VAGINA AND VULVA

This is a fairly common complication which occurs during the birth of a large or abnormally presented foal, especially in maiden mares or mares with a small pelvis. Laceration and tearing of the vulva will occur if a stitched mare is not cut enough to allow the foal to be born.

Damage by the foal as it leaves the birth canal will soon be noticeable: the lips of the vulva swell, appearing congested and painful. The vaginal wall will also be reddened and somewhat inflamed. In this case examination of the mare by your veterinary surgeon is a wise precaution, to help prevent further trouble later on.

The risk of infection to the damaged tissue will be increased and this could lead to a uterine infection and cervicitis, possibly resulting in an infertile mare.

Mares which have sustained bruising and laceration at foaling should not be covered at the foaling heat, but should be left to some future date when they have healed sufficiently. These mares often show signs of an acute and sometimes profuse vaginal discharge, but with veterinary treatment this usually clears up quite soon and the mare will probably then be ready to serve next time she comes in season.

EQUINE VIRUS ABORTION – RHINOPNEUMONITIS VIRUS ABORTION

This disease is endemic in America but sporadic outbreaks do occur in the British Isles almost every year. Vaccination of all foaling mares is routine practice on most well-run studs in the States.

The virus is identical with that causing 'snotty noses'

in young horses — a condition usually seen during the autumn and winter. Abortions occur in infected mares between the fifth and eleventh months of gestation, with most occurring during the ninth and tenth months. Affected foals carried to full term are usually born dying or in a very weak state. The incubation period appears to be between twenty and ninety days.

The following revised recommendations on rhinopneumonitis virus abortion are reproduced by kind permission of the Thoroughbred Breeders' Association:

1. *Characteristics of the disease*
Rhinopneumonitis is caused by a herpes virus — equine herpes virus 1 (EHV1). It is readily destroyed outside the body by heat and disinfectants, but if cleaning and disinfection are not adequate the virus may survive for several weeks.

The virus can cause a mild fever, coughing, nasal discharge and other signs of respiratory disease, usually in yearlings. The respiratory form of the disease occurs most frequently in autumn and winter. However, infection of the respiratory tract and excretion of virus may occur without clinical signs of illness.

Some strains of virus cause abortion in mares late in pregnancy, usually between eight and eleven months of gestation. Aborted foetuses, membranes and fluids are a dangerous source of infection but mares may also acquire infection from horses and foals excreting the virulent strains of virus from the respiratory tract. Abortion can occur from two weeks to several months after infection. Occasionally, infected mares may become inco-ordinated or paralysed. It is suggested prolonged transport and other types of stress during late pregnancy may result in infection of the foetus.

Frequently infected foals are born alive, but weak, with jaundice or breathing difficulty. They are highly infectious. Usually death occurs in one—two days.

After virus abortion or the birth of an infected foal, the virus does not appear to persist long in the mare's genital tract.

The disease can be diagnosed by post-mortem examination of the foetus or foal. Confirmation of the diagnosis may be provided by virological examination of the tissues. This usually requires one—two weeks.

2. *Important points in the control of the spread of equine rhinopneumonitis abortion*
    (i)   Wherever practicable, mares should be foaled at home and sent to the stallion with a healthy foal at foot because mares with a healthy foal at foot are most unlikely to be excreting rhinopneumonitis virus. For practical purposes

this avoids the risk of spread of rhinopneumonitis-induced abortions.

(ii) Where this is not possible, the best solution would be that at least one month before a mare foals, she should be sent to the stud where the stallion is standing and put into group isolation at the stud.

Mares in late pregnancy which have come from a sale or from abroad, constitute a risk to the stud where the mare is sent and should be separately isolated.

(iii) Weaned foals and yearlings should be segregated from pregnant mares as far as possible. If handled by the same attendant the work schedule should be arranged to allow the mares to be handled first.

Present evidence indicates that no harm is done if a mare is covered one month after she has had an abortion caused by rhinopneumonitis virus.

3. *Control procedures*

(i) An inactivated or killed rhinopneumonitis vaccine is now available and mare owners should consult their veterinary surgeon about its use and regarding a suitable vaccination programme. Vaccination programmes formulated to provide protection against rhino-pneumonitis virus abortion should not by themselves be allowed to engender a false sense of security. It must be stressed that vaccination cannot replace sound management practices which should be arranged in consultation with the stud farm veterinary surgeon.

(ii) It is essential that cleansing and disinfection of horse conveyances are carried out in accordance with regulations of the Ministry of Agriculture, Fisheries & Food.

(iii) Isolation facilities should be provided at all studs to prevent the spread of rhinopneumonitis and other infections. The design and construction should be undertaken in consultation with the stud's veterinary surgeon, who should also be consulted about the precautions to be taken to prevent the spread of infection by the attendant.

4. *Action to be taken when abortion occurs*

(i) Action should be taken in the case of any abortion, or of a foal being born dead, or dying within seven days of birth even if the mare concerned is believed to have been vaccinated against rhinopneumonitis. The stud's veterinary surgeon must be notified immediately and arrangements made, on his advice, for the foetus and membranes or carcase to be sent, complete, to a recognised centre for post-mortem examination. If the foal is alive but possibly infected a diagnosis may be made by virological examination of nasopharyngeal swabs taken from the foal. These should be submitted to an appropiate laboratory as soon as possible.

(ii) The mare should be put into strict isolation pending result of the examination, all bedding and traces of foaling should be burnt, the mare washed down with a disinfectant, such as one of the Iodophor group, and the box thoroughly cleansed using water under pressure or steam according to the instructions of the stud's veterinary surgeon.

(iii) If rhinopneumonitis is diagnosed at a stud, the appropriate Breeders' Association should be notified immediately. At the same time owners or their agents with mares at the stud, or due to send mares there, should be informed. Those studs to which mares from the infected premises have been, or are to be, sent should also be informed.

Notification of all cases to the Breeders' Association is of the utmost importance to the long-term interest of all owners of mares. No stigma is attached to an abortion attributable to rhinopneumonitis on a stud, but failure to notify the disease leads directly to the spread of the infection to the detriment of all owners and their horses, especially mare owners.

(iv) Any foal which is ill at birth or becomes unwell within seven days ot birth should be carefully examined by a veterinary surgeon.

5. *Subsequent action*

(i) Providing there is no sign of infection *at home*, barren and maiden mares, and mares which have foaled *at home* and produced normal, healthly foals, can be accepted on the stud in question at any time. Although foals should be ten days old before being exposed to the virus, as there is some risk of pneumonia.

(ii) These mares (enumerated in (i) above) can leave the stud at which the abortion has occurred after one month from the date of the last abortion at the stud, provided they are isolated from in-foal mares at their home stud for two months.

(iii) The stud's own mares can visit other studs at any time after one month from the date of the last abortion, provided they can be isolated for two months from in-foal mares at the stud they are visiting. Full agreement should be reached between the home stud and the visiting stud and their respective veterinary surgeons on these arrangements.

(iv) Mares returned from studs where rhinopneumonitis virus abortion has occurred during the previous season should be foaled in isolation at home.

CHECK-LIST FOR ACTION IN CASE OF AN ABORTION

A. *When an abortion occurs, or a foal is born dead, or dies within seven days of birth*
1. Notify the stud's veterinary surgeon and obtain his instructions . . . .
2. Send the foetus and its membranes or carcase for post-mortem examination as instructed by the veterinary surgeon . . . .
3. Put the mare in strict isolation . . . .
4. If the foal is alive but possibly infected, isolate the mare and foal. The attendant should have no contact with pregnant mares . . . .
5. Destroy bedding, clean and disinfect premises and vehicles under veterinary supervision . . . .
6. Notify owners (or their agents) due to send mares to the stud . . . .

B. *If rhinopneumonitis infection is confirmed notify*
1. The appropriate Breeders' Association
   (a) by telephone . . . .
   (b) in writing . . . .
2. Owners (or their agents)
   (a) with mares already at the stud . . . .
   (b) due to send mares to the stud . . . .
3. Other studs
   (a) to which mares from the infected premises have been sent . . . .
   (b) to which mares from the infected premises are to be sent . . . .

(A copy of this check-list must be displayed in the tack room or somewhere the personnel can see it easily).

An outbreak of the disease occurred on a stud in the Newmarket area in 1979, resulting in the deaths of ten mares following paralysis. A similar outbreak occured at the Lippizzaner stud in 1983. This is a very rare occurrence. Stallions, barren mares or even geldings, as well as foaling mares, can be affected. Treatment with corticosteroids is recommended.

HAEMORRHAGE OF THE UTERINE ARTERIES

This condition usually occurs within a few hours of foaling. In the majority of cases there is no sign of blood escaping from the vulva. When the mare is suffering from an internal haemorrhage and is losing a large quantity of blood, the mucous membranes (i.e. the gums and eyes) become pale and anaemic-looking. The mare may sweat profusely and show signs of colicky pain. As she gets weaker, she will be unable to stand and may finally become unconscious.

In some cases the bleeding will stop automatically but in more severe cases of haemorrhage the outlook is very grave; however, blood transfusions can sometimes be given, but in the case of severe haemorrhage these will probably be of little use.

This condition is more common in older mares (i.e. ten years and over) which have had several foals, and more so those with dipped backs and dropped bellies. It is probably due to stretching of the uterine ligaments and arteries, the actual haemorrhage being caused by rupture of one of the arteries to the uterus due to damage in the wall of the vessel.

INFLAMMATION OF THE UTERUS

This condition may be either acute or chronic. In the chronic form the mare becomes permanently infertile, while in the more acute form the mare may recover with time.

When acute inflammation occurs immediately after foaling it is usually due to the introduction of septic bacteria (germs) into the uterus during foaling. These organisms are usually conveyed by the unwashed hands and arms of the attendants or from unsterile ropes and instruments used to assist the foaling. It can also arise from a portion of retained after-birth; this is usually a piece of the non-pregnant horn (see plate 16) which remains attached inside the mare. Occasionally, due to the great weight of the portion lying outside the mare, the non-pregnant horn is torn through and is thus retained.

This is a very severe condition and if left untreated is fatal. Your veterinary surgeon should be called in immediately the symptoms are observed. The mare becomes very ill and loses all interest in her foal, usually within the first two days after foaling. There is usually a bloodstained grey discharge from the vulva, which will soil her tail. In very bad cases the mare may appear tucked up and stand around with her back arched, in obvious pain.

As far as inflammation is concerned, prevention is better than cure. Extreme cleanliness should be observed at foaling time. If it becomes necessary to insert your hands and arms into the vagina during foaling you must make sure that they are scrubbed clean in a disinfectant, that your nails are short and free from dirt and that disposable plastic gloves are worn. All instruments and ropes used for foaling must be sterilised, preferably by boiling for ten minutes, after which they should be left covered in the container in which they were boiled until needed. Loose-boxes to be used for foaling must be swept clean of dust and cobwebs and the straw must be the cleanest available.

MASTITIS

As compared with the cow this condition is comparatively rare in the mare, but nevertheless it does occur occasionally.

Mastitis is inflammation of the udder tissue due to a bacterial infection. It can be brought about by wounds or scratches on the teats and udder which become contaminated by flies or from lying on dirty bedding. It can also occur on occasion as a secondary infection to a disease such as strangles.

The mare's udder will appear hot, swollen and quite hard to the touch, due to the extreme pain in this region. She will probably not let her foal suckle and may attempt to bite or kick it should it go near her udder. Her temperature, respiration and pulse will be higher than normal; she may go off her food and be subject to occasional fits of shivering. If some milk is drawn from the udder, it will appear thin and watery and will contain several clots.

The mare will probably tend to walk stiffly with the hind leg of the affected side. In bad cases the swelling may extend forward from the udder, right along the entire length of the stomach to the front legs.

It is imperative that your veterinary surgeon should be called as soon as the first symptoms are noticed. If neglected, this condition will lead to the complete destruction of udder function on the affected side.

Mastitis is very occasionally found in filly foals and maiden or barren mares but most often occurs at weaning time or when mares with foals at foot are first turned on to very rich pasture, which produces a flush of milk.

RECTO-VAGINAL FISTULA

This condition is fortunately uncommon but may occur when, due to abnormal presentation of the foal in the birth canal, one foot is pushed through the wall of the vagina and into the rectum. It is most common in first-time foaling mares.

The symptoms in individual mares will depend on the extent and position of the injury. In some cases the foot may be seen protruding through the rectum.

Where possible, the foal should be pushed back inside the mare. In any case your veterinary surgeon should be called in immediately. If the position of the foal cannot be rectified the mare should be got on her feet and walked round until the veterinary surgeon arrives, to prevent her from straining further.

In most cases an operation to repair the tear can be performed with success.

RETENTION OF THE AFTER-BIRTH

In this case the after-birth is not expelled in the normal way by the secondary birth pains, after foaling, but remains attached inside the mare. It must be removed within twelve hours of foaling or infection may set in, leading to inflammation and possible general septicaemia with fatal results. In this respect the mare is far more susceptible than the cow, which can often be left for days with no apparent ill effects.

On no account should an owner attempt to remove the after-birth himself, as traction on the visible part of the membranes would almost certainly lead to some of the attached portions tearing inside the mare. The retained portions would decompose, leading to an acute infection and the probable death of the mare.

As a general rule of thumb if, when the mare has foaled during the night, the membranes are still retained first thing in the morning, your veterinary surgeon should be notified immediately before he goes out on his rounds. When he arrives he will need a clean bucket of warm water, some soap and a clean towel.

UTERINE PROLAPSE

This can occur any time up to about twenty-four hours after foaling. In some cases only the vagina appears outside the walls of the vulva but in other cases the whole of the uterus invaginates and appears hanging downwards as a large fleshy mass often reaching as far as the hocks. This condition can follow an easy foaling and is probably more common in mares with high-set tails and slack ligaments round the vulva.

A partial prolapse of the uterus can occur internally when one of the horns turns in on itself; in this case the mare will show very definite signs of colic, she will sweat profusely and get down and roll in an effort to ease the pain. There will be no other visible sign that there is anything wrong. Your veterinary surgeon should, however, be called in immediately.

If the whole of the uterus is prolapsed, it must be kept clean and warm until the veterinary surgeon arrives, and great care should be taken to prevent its getting damaged in any way. The easiest way to do this is to place the mass on a clean sheet and remove all the

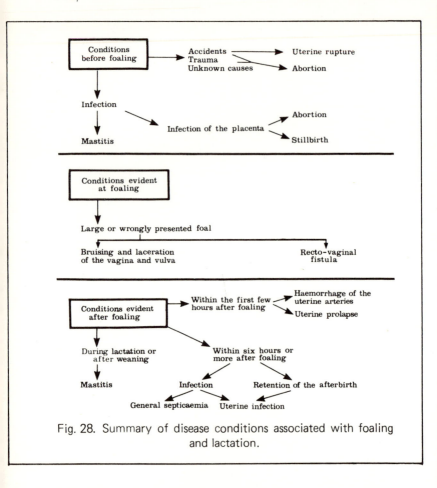

Fig. 28. Summary of disease conditions associated with foaling and lactation.

particles of straw which may be adhering to it. Clean warm water may be sprinkled over the uterus to prevent it from drying while you await the arrival of the veterinary surgeon, as the longer it remains outside the body the more swollen and dry it becomes.

Mares have recovered after a uterine prolapse but generally speaking unless treatment is prompt the outlook is grave.

UTERINE RUPTURE

This is a very serious, although fortunately rare, condition which can occur at any time during pregnancy but especially at foaling time. It is associated with a severe kick, blow or more commonly with a bad fall. It can occur during foaling due to the foal putting one of its feet through the wall of the uterus.

Rupture of the uterus invariably means that the animal will die unless an abdominal operation can be performed to close the wound, often a very difficult procedure.

16 Contagious Equine Metritis

Contagious equine metritis is a highly infectious venereal disease which usually causes infertility. It was first publicised at the English National Stud in 1977, and is referred to as CEM for short. CEM was also confirmed among Thoroughbreds in Ireland, Australia and other studs in England in 1977 and in the State of Kentucky, USA, in 1978. The disease was confirmed among Thoroughbreds and other breeds in France in 1978. The causal organism is a new species of the genus Haemophilus and is known as *Haemophilus equigenitalis*. These bacteria do not grow under normal aerobic culturing procedures in the laboratory but require special media in the presence of hydrogen and carbon dioxide (micro-aerophilic conditions). Most, but not all, infected mares have a copious grey/translucent discharge from the vulva.

Mares and stallions appear to recover from CEM without any ill effects, although some mares become carriers and continue to harbour the organism, so are capable of infecting clean stallions and mares. It has been found that carrier mares seem to harbour the organism in the area around the clitoris, so swabs are taken from the clitoral fossa and sinuses as well as from the lining of the uterus. Experience has shown that the best results are obtained when uterine swabs are taken during the early part of the heat period.

'It is an ill wind that blows nobody any good,' so the saying goes, and this is probably true for CEM. Since the disease was first discovered in 1977, veterinary surgeons, stud managers, owners and stud hands have all been made doubly aware of the necessity for hygiene when handling breeding stock, which has led to improved standards on stud farms in the British Isles. A code of practice was drawn up in 1978 and since amended, including for the 1984 covering season. It makes the following definitions and recommendations:

1. Mares have been divided into two categories, high and low risk, for the purpose of bacteriological examinations. The recommendations are the minimum requirements for the examination of mares and stallions *and it is emphasised that owners should consult their veterinary advisers concerning the implementation of the Code.* As more information becomes available amendments to the Code may be introduced.

DEFINITIONS

2. The definitions used are as follows:

(A)   *High-risk mare*
   (i)   A mare from which the CEM organism was isolated on a previous occasion.
   (ii) A mare covered by a stallion which transmitted CEM in the previous year.
   (iii) All mares arriving from countries other than France, Ireland and the United Kingdom or covered by stallions resident in countries other than France, Ireland and the United Kingdom in the previous year.

Information with regard to the identity of high-risk mares may be obtained from the Thoroughbred Breeders' Association, Stanstead House, Newmarket, Suffolk.

(B)   *Low-risk mare*
   A mare not defined as 'high risk' above.

SWABS
   (i)   *Mares*
      (a) *Endometrial swab*
      A swab taken during early oestrus from the lining of the uterus.
      (b) *Clitoral swab*
      A swab taken from the clitoral fossa including the clitoral sinuses, using a swab suitable to penetrate the clitoral sinuses.
   (ii) *Stallions*
      (a) A sample of pre-ejaculatory fluid.
      (b) A swab taken from the penile sheath.
      (c) A swab taken from the urethra.
      (d) A swab taken from the urethral fossa.

All swabs should be immediately despatched to a designated laboratory and be cultured under both microaerophilic and aerobic conditions so that the presence of the CEMO and other pathogens may be identified and subsequently reported. Microaerophilic cultures for CEM should be incubated for a minimum of six days although the laboratory may issue a preliminary report at four days if requested.

RECOMMENDATIONS TO MARE OWNERS

ALL MARES
3. Owners should obtain a detailed history of their mares whilst at stud in the previous three years and the information should be entered on Certificate A (see para 12). This should include the results of positive bacterlological tests undertaken including tests for the CEMO since 1977. The results of tests carried out just prior to the time the mare is moved to the stallion stud farm should be entered on Certificate B (see para 12) which should be completed by the vertinary surgeon who examined the mare. The stallion stud farm should receive the Mare Certificate before the mare is sent to the stallion stud farm.

REQUIREMENTS
4. (A) *High-risk mares*
  (i) Prior to going to the stallion stud farm, one negative clitoral swab.
  (ii) On arrival at the stallion stud farm, one negative clitoral swab.
  (iii) On the stallion stud farm, one negative set of endometrial and clitoral swabs should be taken during the oestrus prior to covering.

(B) *Low-risk mares*
  (i) One negative clitoral swab. This may be taken either at the stallion stud farm or on the home stud farm providing there is agreement with the stallion stud manager.

(C) *All mares*
  (i) On the stallion stud farm prior to covering an endometrial swab should be taken during oestrus and cultured aerobically to identify other bacteria capable of causing venereal disease.

ABORTION
5. If abortion occurs the foetus and placenta should be sent to a laboratory which regularly carries out equine post-mortem examinations and the mare kept in isolation until it is shown to be free from infection.

IN-FOAL HIGH-RISK MARES
6. In-foal high-risk mares should be foaled in isolation with hygienic disposal of the placenta. In addition to the recommendations in 4A, a cervical swab should be taken at foaling or within 12 hours of foaling. Foals born to these mares should be swabbed on 3 occasions at intervals of not less than 7 days before they reach 3 months of age. Filly foals should be swabbed in the clitoral fossa and colt foals should be swabbed inside the penile sheath and around the tip of the penis.

RECOMMENDATIONS TO STALLION OWNERS

ALL STALLIONS AND TEASERS
7. Stallions and teasers should be subjected to a bacteriological examination of the external genital organs by a veterinary surgeon. The examination should be carried out after 1st January but before the start of the covering season. A set of swabs should be taken on two separate occasions at intervals of not less than seven days and the results entered by the veterinary surgeon on Certificate C (see para 12). As an additional precaution the first mares covered by a first season stallion should be screened for CEM. The stallion owner or stud manager will then be able to inform owners of mares booked to a particular stallion of the results of the tests.

HYGIENE
8. It is important that all stud personnel realise the highly contagious nature of CEM. They should only handle the external genitalia of mares and stallions when wearing disposable gloves.

WALKING-IN MARES
9. On the first occasion before each mare is walked-in a clitoral swab should have been taken not more than 30 days before the mare goes to the stallion stud farm. An aerobic endometrial swab should also be taken during oestrus. Additional swabbing should be at the discretion of the veterinary surgeon responsible to the stallion stud farm who will liaise with the veterinary surgeon responsible for the mare. If CEM is confirmed on the boarding stud farm where the mare resides or on the stallion stud farm which it has visited, no mare should be moved from the boarding stud farm until all mares at the boarding stud farm have been swabbed with negative results.

THE EXAMINATION OF MARES AFTER COVERING HAS COMMENCED
10. If in the opinion of the attending veterinary surgeon an abnormal pattern of returns to service develops it is imperative that a full set of swabs is obtained and cultured under aerobic and microaerophilic conditions.

OUTBREAK OF CEM

ACTION
11. The action to be taken when CEM is confirmed is as follows:
(a) *In mares prior to covering*
    (i) Isolate infected mares and treat as advised by the attending veterinary surgeon.
    (ii) Notify all owners of mares booked to the stallion(s) including those which might have already left the stallion stud farm.

       (iii) Nofity the Thoroughbred Breeders Association, Stanstead House, Newmarket, Suffolk CB8 9AJ. Tel: Newmarket (0638) 61321.

(b)   *Mares and stallions after covering*

    Apply (i), (ii) and (iii) as above and in addition

       (iv) Cease covering by the stallion, which should be swabbed and treated as advised. Covering should not be resumed until the stallion has had three negative sets of swabs taken at intervals of not less than two days.

       (v) Check and deal with all mares which are implicated in the outbreak as advised by the veterinary surgeon.

EXPORT CERTIFICATION

11.   The Ministry of Agriculture, Fisheries and Food wish to re-emphasise that the ability of local veterinary inspectors to issue certification will depend on the keeping and retention of accurate records, and upon the full implementation of the Code of Practice, including such recommendations as may be made in the future.

CERTIFICATION

12.   Certificates may be obtained from either the Secretary, Thoroughbred Breeders Association, Stanstead House, Newmarket, Suffolk or Irish Thoroughbred Breeders' Association, Johnstown, Naas, Co. Kildare on receipt of a stamped addressed envelope.

(The above Code of Practice is reproduced by kind permission of the Horserace Betting Levy Board.)

APPENDIX 1

Samples for Laboratory Tests

There are many instances when you may require to take a sample from your horse to give to your veterinary surgeon for testing. In these cases it is very useful to know the amount of sample required by the laboratory for an accurate result.

Only samples which are usually collected by the owner for the veterinary surgeon are listed and not those collected by the veterinary surgeon himself.

FAECES SAMPLES

1 Worm egg count for red worm, roundworm, tapeworm or liver fluke:
Enough sample to fill an Equizole powder tin or universal bottle.

2 Lungworm larvae count:
The minimum quantity required for a satisfactory determination is a half-pound honey jar absolutely full; a lesser amount gives an inaccurate result.

3 Occult blood test:
Minimum quantity is a universal bottle full; the container must be sterile.

4 Bacteriology:
Minimum quantity as above and collected into a sterile receptacle; it must be a really fresh sample, preferably taken before it reaches the ground.

As far as faeces samples for worm counts are concerned, they should be put into an airtight container such as a sealed tin, wax pot or polythene bag – they must not be allowed to dry out. They should be placed somewhere cool, out of direct sunlight, otherwise the eggs may hatch out and a false result will be obtained. All samples must be from absolutely fresh droppings.

URINE SAMPLES

1 For pregnancy testing:
A minimum quantity of a little under half a universal bottle full is quite sufficient but a larger quantity should be sent wherever possible. The container need not be sterile but must be absolutely watertight.

2 For routine examinations:
The minimum quantity is a one-pound honey jar full; the container should be absolutely clean and dry and preferably should be sterile.

3 For bacteriological examinations:
The minimum quantity is a universal bottle full; the container must be sterile and the sample should be collected straight into the bottle from midstream of the urination.

SKIN SCRAPINGS

These are usually taken for the evidence of mites, ringworm and dermatophalus — lice can be seen with the naked eye. The sample must be plucked or scraped from the actual site of the infection and incorporate as much skin tissue and hair roots as possible and not just the hairs themselves. On no account should the hairs be cut or clipped off. Sufficient sample to fill a matchbox, if possible, would be ideal for the tests.

As ringworm is very contagious to man, a pair of rubber gloves should be worn when collecting the sample.

MILK SAMPLES

These are usually collected where inflammation and disease of the udder are suspected. The sample must be collected by a sterile technique into a sterile container, minimum one universal bottle full. Using a piece of cotton wool soaked in methylated spirits swab your hands and the mare's teats, and draw some milk from the affected teat or teats into the sterile bottle. Give the sample to your veterinary surgeon as soon as possible.

FOETUS AND MEMBRANES

Should you have the misfortune to own a mare which slips or aborts her foal and you happen to find the foetus and membranes, these should be collected into a clean polythene bag and given to your veterinary surgeon as soon as possible. On no account should the foetus and membranes be washed, not even with plain water let alone disinfectant. Wherever possible make sure that both foetus and membranes are included.

(A universal bottle is a small glass bottle with a metal screw top and measures 8.5 cm high, and has a diameter of 2.75 cm.)

APPENDIX 2

Further Reading

The Practice of Equine Stud Medicine, Rossdale and Ricketts, Baillere and Tindall, 1974.

Stallion Management, A.C. Leighton Hardman, Pelham Books, 1979.

Young Horse Management, A.C. Leighton Hardman, Pelham Books, 1976.

A Guide to Feeding Horses and Ponies, A.C. Leighton Hardman, Pelham Books, 1977.

Equine Nutrition, A.C. Leighton Hardman, Pelham Books, 1980.

APPENDIX 3

Useful Addresses

BORD na gCAPALL (Irish Horse Board),
Blessington Road,
Tallaght,
Co. Dublin,
Ireland

BRITISH HORSE SOCIETY,
British Equestrian Centre,
Stoneleigh,
Kenilworth,
Warwicks CV8 2LR

HUNTERS' IMPROVEMENT SOCIETY,
G.W. Evans Esq,
National Westminster Bank Chambers,
8 Market Street,
Westerham, Kent

NATIONAL FOALING BANK,
Miss Johanna Vardon,
Meretown Stud,
Newport, Salop
Tel: (0952) 811234

THOROUGHBRED BREEDERS' ASSOCIATION,
S.G. Sheppard Esq,
Stanstead House,
Newmarket,
Suffolk CB8 9AJ

WEATHERBYS,
Sanders Road,
Wellingborough,
Northants NN8 4BX

Index